gear
equity
mainstream : courant dominant

Finance and the Small Firm

This edited collection is based upon work carried out under the ESRC's Small Business Programme. It examines key issues in the financing of small businesses. The starting premise is that there are imperfections in the market for the provision of finance for small firms. Excessive dependence upon short-term facilities such as loans and overdrafts has inhibited more long-standing relationships between small firms and providers of finance which would be more mutually beneficial.

The contributors call for changes which would solve this current imbalance. They suggest that small firms need to become more 'professional' in their financial management, with owners more prepared to 'share' equity. Financial institutions need to rely less on collateral and concentrate more on making informed judgements about the qualities of each business.

In this volume small firms are not treated as one indistinguishable mass but are broken down into specific types: the micro firm; the 'high-tech' small firm and the small firm owned by an individual from an ethnic minority. This treatment is unique.

Written by experts in small business, this volume will be of great interest to researchers in small business, students of industrial economics, business finance and management studies and by government policy advisers and financial advisers.

Alan Hughes is Director of the ESRC Centre for Business Research at the Department of Applied Economics, University of Cambridge. **David J. Storey** is Director of the SME Centre, Warwick University Business School.

Routledge Small Business Series
Edited by David J. Storey

Finance and the Small Firm

Edited by A. Hughes
and D.J. Storey

the department for Enterprise

London and New York

First published 1994
by Routledge   ʰ20ᶫ 658·1592 FIN
11 New Fetter Lane, London EC4P 4EE

Simultaneously published in the USA and Canada
by Routledge
29 West 35th Street, New York, NY 10001

© 1994 A. Hughes and D.J. Storey

Typeset in Times by J&L Composition Ltd, Filey, North Yorkshire
Printed and bound in Great Britain by
Biddles Ltd, Guildford and King's Lynn

British Library Cataloguing in Publication Data
A catalogue record for this book is available from the British Library

Library of Congress Cataloging in Publication Data
Finance and the small firm / edited by A. Hughes and D.J. Storey.
    p.   cm.—(Small business series)
    Includes bibliographical references and index.
    ISBN 0–415–10036–4 ✓
    1. Small business—Great Britain—Finance. 2. High technology
industries—Great Britain—Finance. 3. Minority business
enterprises—Great Britain—Finance. 4. Venture capital—Great
Britain. 5. Small business—Great Britain—Finance. I. Hughes, A.,
1946– . II. Storey, D.J. III. Series: Small business series
(London, England)
HG4027.7.F55   1994   94–11409
658.15′92—dc20   CIP
ISBN 0–415–10036–4

Contents

Figures

Tables

Contributors

Giles Barrett is a Part-time Lecturer and Research Student at the School of Social Science at Liverpool John Moores University.

Andy Cosh is Lecturer in Management Economics and Finance at the Judge Institute of Management Studies and Assistant Director of the ESRC Centre for Business Research at the University of Cambridge.

Judith Freedman is Senior Lecturer in Law at the London School of Economics.

Sheila Greenfield is Research Fellow in Accounting at the University of Birmingham.

Michael Godwin is a part-time Lecturer in the School of Social Sciences, University of Bath.

Richard Harrison is Professor of Management Development, Centre for Executive Development, Ulster Business School, University of Ulster.

Alan Hughes is Director of the Small Business Research Centre at the Department of Applied Economics, Cambridge University.

Trevor Jones is Reader in Social Geography at the School of Social Science at Liverpool John Moores University.

David McEvoy is Professor of Urban Geography and Director of the School of Science at Liverpool John Moores University.

Colin Mason is Reader in Economic Geography, University of Southampton.

Barry Moore is Assistant Director of Research in the Department of Land Economy and a member of the ESRC Centre for Business Research at the University of Cambridge.

Amanda Nayak is Lecturer in Accounting at the University of Birmingham.

David J. Storey is Director of the SME Centre, Warwick University Business School.

Preface

This volume reports the results of research supported by the Economic and Social Research Council's Small Business Research Programme. Additional support for the programme has been provided by Barclays Bank, the Commission of the European Communities (DG XXIII), the Department of Trade and Industry and the Rural Development Commission. This support is gratefully acknowledged, although the views expressed do not necessarily reflect those of the sponsoring organisations.

At a personal level the research has benefited from the active participation of the co-sponsors, notably John Martin from Barclays Bank, Martin Harvey from the Commission of the European Communities, Keith Lievesley from the Rural Development Commission and Cliff Baker from the Department of Trade and Industry.

Abbreviations

BES	Business Expansion Scheme
BVCA	British Venture Capital Association
CTNs	Confectionary, tobacconists and newsagents
DCF	Discounted cash flow
GPS	Ratio of gross profits to sales
IOD	Institute of Directors
M & A	Mergers and acquisitions
MBI	Management buy-ins
MBO	Management buy-outs
NTBFs	New technology-based firms
POH	Pecking order hypothesis
PRH	Private rented housing
RODNA	Ratio of net profit plus directors' emoluments to total assets
ROE	Rates of return on equity
RONA	Rates of return on net assets
ROTA	Rates of return on total assets
SMART	Special Merit Award for Research and Technology
SME	Small and medium-sized enterprise

1 Introduction: financing small firms

Alan Hughes and David J. Storey[1]

THE BACKGROUND

The idea that problems in the financing of smaller firms have significantly hindered the role they play in the overall performance of the UK economy is deeply rooted. Successive Committees of Inquiry into Small Firms (The 'Bolton' Report HMSO 1971), and into the Functioning of Financial Institutions (The 'Wilson' Report HMSO 1979), identified problems for small and medium-sized independent owner-controlled firms employing less than 200 (or 500) employees (SMEs). The Wilson Committee in its report on SME financing argued for instance that SMEs were relatively risky. They could therefore expect to face higher interest charges or security conditions than larger firms. They concluded nevertheless that excessive bank caution led to smaller, and especially newer, SMEs being rationed in the market for loans and bank finance. They also pointed to a shortage of start-up capital and of equity development capital for fully geared established businesses wishing to expand. These reports led over the years to the introduction of a variety of policy measures designed to tackle 'market failures' in the supply of finance to SMEs, and particular sub-groups of them, such as hi-tech firms, with a specific requirement for risk capital. For instance, in the 1980s the Loan Guarantee Scheme (LGS) was introduced alongside the Business Expansion Scheme (BES). These were designed to tackle respectively the debt and equity sides of the SME financing problem. In the 1990s the Loan Guarantee Scheme has been augmented and although the BES was first wound down but then essentially reintroduced as the Enterprise Investment Scheme, other (very different) schemes such as SMART and SPUR were introduced to support investment in hi-tech and innovative businesses. There was also a substantial increase in the availability of venture capital funding (Hughes 1992).

The apparent resurgence, alongside these changes, of the SME sector in the 1980s, and the identification of contemporary government fiscal policy generally with corporate and personal tax cuts to encourage an 'enterprise culture' might have been expected to reduce the intensity of claims that the SME sector in the UK remains at a disadvantage. Indeed one recent academic survey concluded that 'small firms in Great Britain currently face few difficulties in raising finance for their innovation and investment proposals in the private sector' (HMSO 1991: 17).

This has not however proved to be the final word. The combined impact of high interest rates and prolonged recession in the early 1990s served to highlight the extent to which SMEs continue to rely heavily on short-term finance from banks, to finance their activities. It also revealed the way in which, when times are hard, the smallest businesses, which proliferated most rapidly during the 1980s, and are most dependent upon banks, protest vigorously about the apparently unsatisfactory nature of their relationship with them (Binks *et al.* 1992a, Cowling *et al.* 1991, Bink *et al.* 1992b, Bank of England 1993). Moreover, a recent study by the Advisory Council on Science and Technology of barriers to growth in small firms echoed the Wilson and Bolton Reports' concerns over equity, especially its absence in small amounts for risky projects, and those involving significant innovative components. In this connection it pointed to the much better developed informal venture capital market in the US. It also noted a number of instances in the case studies it conducted in the UK where financing difficulties had led to the sale of otherwise viable independent businesses to larger concerns, with the implication that where foreign purchasers were concerned an opportunity to develop an internationally competitive UK concern had been lost (ACOST 1990).

Moreover other developments in the 1980s led to the emergence of new questions in the debate over small business finance. One of these was the growth of ethnic businesses, in particular those owned and run by members of the Asian community. Here the debate focused on their particular strengths and weaknesses in raising finance in the face of cultural stereotyping, lack of established track records, and racial discrimination on the one hand, and strong informal financial networks on the other (Deakins *et al.* 1992, Ward 1991). At the same time the unprecedented rate of company formation in the 1980s was associated with a debate about the relatively high cost of compliance with company audit and disclosure requirements for small firms compared with their larger counterparts, and the widespread use by the former of the dispensation to submit brief 'modified accounts' granted under the Companies Act of 1981. It also led to a debate about whether the benefits of limited liability were worth the costs incurred for smaller companies in a world in which the pursuit of collateral by banks effectively undermined it, and in which the supposed benefit of company status in gaining access to capital could rarely be taken advantage of because of the equity capital market failures noted above.

THE CHAPTERS

The chapters which make up this book address a number of these issues. They represent the results of research carried out in the period 1989–92 under the Small Business Programme of the Economic and Social Research Council. They are timely in the sense of reporting fieldwork and statistical analysis covering the most recent period in which the financing of smaller firms has re-emerged as a policy issue. At the same time the collection places this period in a wider historical context and brings new light to bear on the issues involved both in terms of theoretical developments and better data than has hitherto been the case.

In their opening chapter Cosh and Hughes (Chapter 2) provide a long-run analysis of trends in SME profitability and financing patterns in the UK. They do this by linking the results of the earlier Wilson and Bolton analyses with two unique databases. The first is a size stratified panel sample of the whole company sector for the years 1977–83. The second is a sample of over 2,000 independent UK SMEs which responded to the Cambridge University Small Business Research Centre National SME survey carried out in 1991 (SBRC 1992). Making use of the insights of modern financial theory they show that the financial structure of smaller businesses compared with larger ones is consistent with a 'pecking order hypothesis' (Myers and Majluf 1984). Here funds are sought in an order which minimizes external interference and ownership dilution by leaving equity till last after retentions and debt have been exhausted. This also has positive signalling qualities in so far as preferring debt to an early issue of equity implies confidence in high returns and an unwillingness to share them with new shareholders. The upshot is that small firms are characterized by a relatively greater reliance on short-term loans and overdrafts and a much smaller reliance on equity finance than are larger firms. This structure is shown to be of long standing, along with the greater significance of trade debt and trade credit in their balance sheets. Cosh and Hughes however point out that the long boom of the 1980s was associated with a substantial increase in short term gearing as small companies found it possible to gain access to bank finance on a large scale. The impact of high interest rates and recession were therefore even more damaging than they might otherwise have been when the long boom ended. They also show that for a sector dependent in the first instance on retained profits to finance growth the 1980s were not historically a particularly auspicious period for SMEs. Whereas in the 1960s and to a lesser extent the 1970s they had recorded somewhat higher returns than larger firms, by the 1980s the position was reversed. Although the quality of the data requires the exercise of some caution it appears that small firms suffered particularly badly from the impact of the recession of the early 1980s. One important conclusion to emerge from this work is that the financial structure of SMEs reflects the wishes and strategies of their owners as much as constraints placed upon them by suppliers of finance. It is therefore necessary to probe qualitatively beyond the financial structure itself to address issues of financial constraints more directly. Here case study evidence reviewed by the authors supports theoretical arguments for some market failures on the loan and equity fronts for rapidly expanding or innovative firms, as well as the relative paucity of informal venture capital in the UK. In the latter case in particular their survey data supports the case study material. Informal funds (i.e. supplied by private individuals who were not working partners or owner-directors[2]) amounted on average to only 1.7 per cent of total new funds raised by their sample of around 2,000 SMEs in the period 1987–90, whilst formal venture capital averaged around 2.9 per cent (SBRC 1992). The contrast with the United States is striking. In that country informal venture capital is twice as important in quantitative terms as institutionalized venture capital (Wetzel 1987). However, the Cambridge Survey also revealed that informal capital was at its most significant amongst micro-firms employing less than ten people. In those firms it accounted for 3 per cent of new funds compared with 1 per cent from formal venture sources.

The nature of this type of informal finance is explored in the chapter by Mason and Harrison (Chapter 3). One of their main contributions, in the first systematic attempt to survey this sector in the UK, is to identify the characteristics of UK informal investors ('business angels') and their investment practices. The problems of surveying this sector are formidable and are well set out in their chapter which reports the results of a postal questionnaire sample of seventy-eight investors supported by eight face to face interviews. The small sample sizes reflect the paucity of systematic information to serve as a sampling frame as well as the reticence of 'angels' (shown to be predominantly middle-aged businessmen with a continuing interest in other business(es) they have founded) to reveal themselves to the public gaze. Mason and Harrison's analysis confirms the role played by 'business angels' in supplying relatively small sums (median value of around £22,000) to smaller businesses suggested by the Cambridge Survey results. Their detailed analysis shows that this reflects the relatively modest incomes of business angels (only 16 per cent reported earnings above £100,000), their small individual investment portfolios (75 per cent had invested under £50,000 in total in the previous three years, with only 12 per cent of individual investments in amounts over £50,000), and their reluctance to take part in syndicated deals (66 per cent always or usually acted independently). Nevertheless, the highly skewed nature of both income and wealth in the Mason and Harrison sample meant that in total their eighty-six respondents had available investible funds amounting to nearly £10 million.[3] In addition to their potential and actual role as financiers, the strong entrepreneurial and managerial background of the typical investor in their sample meant that they typically played a 'hands-on' role as minority shareholders, joining the board in around 27 per cent of cases, and playing a purely passive role in only 20 per cent. Whilst this may be beneficial in cases where injections of both expertise and finance are required it may have costs in terms of conflicts over strategy and control where they are not (Harrison and Mason 1992). Moreover, given that over 50 per cent of the sample expected to 'exit' from their investment within three to five years, the implications from the point of view of growing medium-sized independent companies are not necessarily promising. Indeed, as Mason and Harrison note, UK angels are less 'patient' than their US counterparts, and more likely to expect to sell to 'outsiders' via merger or stock market flotation than to sell their shares to 'insiders'. From a policy point of view, what emerges most strongly is that if this sector of the financial market is to play a more significant role, a greater degree of information exchange about potential investors and investment opportunities is required. Thus Mason and Harrison show the essentially *ad hoc* nature of much business angel search and referral activity. The essentially individualistic nature of the informal investment activity which their survey reveals leads the authors to conclude that one possible avenue for policy intervention is to be found in the promotion, especially at a local and regional level, of information brokerage, either through the TEC system or the 'Business Links' being introduced by the DTI.[4]

The informal investors investigated by Mason and Harrison are primarily interested in high growth potential businesses, many but not all of which are to found in sectors where innovation-intensive activity offers the prospects of spectacular breakthroughs. The financial problems facing these New Technology Based Firms (NTBFs) has been of particular policy concern, not least because findings such as those of Pavitt *et al.* (1987) have suggested that the share of innovations accounted for by smaller firms since the 1950s in the UK has outstripped the contribution which might have been expected given their shares of output or employment.

It has been suggested in the reviews by Barber *et al.* (1989) and by ACOST (1990) that, although NTBFs play a crucial role in the economy, they find it disproportionately difficult to obtain appropriate financing from the institutions. This is because such firms are perceived to be particularly risky for several reasons. The first is that they are frequently attempting to introduce products and processes which are new to, and untested in, the market. Secondly, the firms are often in industries where rapid developments make existing technology obsolete, and thirdly, they are often in businesses which have only a single product. Finally the businesses are often owned and managed by individuals with stronger technical than business skills. For all these reasons it is not surprising that studies such as those by Oakey (1984) in the UK and Roberts (1991) in the United States have suggested that bank finance is significantly less important at start up for NTBFs than is the case for conventional small businesses. Even so, these results contrast with those of Monck *et al.* (1988) which indicate little difference in the ways in which high-technology firms are initially funded, compared with small businesses as a whole.

The chapter by Moore in this volume (Chapter 4) provides a detailed analysis of the issue. He takes 292 high-technology small companies and compares them with more than 1,700 smaller companies which are not in the high technology sector. Both groups are asked the same questions about the extent to which a range of factors have constrained the growth of their business. For both high-technology *and* conventional firms, finance constraints are seen to be the most important. In a multiple regression analysis Moore demonstrates that the businesses which are most likely to report financial constraints are those which are young, in the manufacturing sector, have below average profitability and are smaller. Holding these and other variables constant, Moore can find no evidence that the high-technology business is more likely to experience difficulties in obtaining finance than its counterpart in the conventional sectors. Indeed it appears to be the case that the small high-technology firm is, if anything, more likely to be in receipt of venture capital and support from private individuals (angels) than is the case for the conventional firm.

The second special case is the financing of ethnic or non-white businesses. The central issue here is whether there is evidence of the owners of these businesses being denied access to loan or equity capital on racial grounds and whether the mode of financing influences the performance of the ethnic business sector.

The major research finding on this topic in the United States has been provided by Bates (1991). Bates' work finds that in the United States start-up businesses established by blacks tended to receive smaller loans from banks than white-owned start ups. Secondly he found that black-owned firms were under-capitalized, compared with white-owned firms.

Bates hypothesizes that commercial banks are more likely to lend to individuals with more human capital, more equity and with demographic traits that are associated positively with business viability. In this context human capital is likely to be reflected in the level of education of the individuals, their age, whether or not they have previous managerial experience and family small business background. The first key result which Bates generates is that, even when these factors are taken into account, it is still the case that the loans made to white business start ups exceed those made to black business start ups.

Bates then shows that it is *lower* levels of capitalization which are associated with higher risk of business failure. From this he infers that at least part of the observed higher failure rate of black-owned firms reflects this lower capitalization and that, most significantly, their failure rates would be no different from white-owned start ups if they received a similar level of external loan.

Research on the financing of ethnic businesses in the United Kingdom has not made use of the huge databases or the statistically sophisticated techniques which are employed by Bates. Nevertheless the work by Ward and Reeves (1980), Wilson and Stanworth (1987), and Deakins *et al.* (1992), provides helpful insights into this topic in the UK. In this volume, Jones, McEvoy and Barrett (Chapter 5) review the financing of both white and non-white businesses. They show that the most striking difference between the ethnic groups, in terms of mode of financing, is not between whites and non-whites, but between Asians and Afro-Caribbeans. They show that Afro-Caribbeans are perhaps twice as likely to experience problems obtaining bank loans as white applicants. Secondly, they show that the Asian business is significantly more likely to have been established using loans from family or friends than is the case either for white or Afro-Caribbean businesses. Thirdly, they point to recent, but highly significant, changes in the ways in which Asian businesses are being financed. They suggest that, if anything, UK banks now look *more* favourably upon the Asian business owner, than upon the white business owner. However, in a very interesting analysis of those businesses established by Asians which have been funded 100 per cent by the bank, the authors show these are much more likely to have been started by British-born Asians than those born overseas, and more likely to have been established by Asians with higher educational qualifications.

It is in their comments upon how matters are likely to change that the authors offer important policy perspectives. Although there is now clearly a much greater willingness of banks to support Asian-owned businesses, the evidence is that bank support is more difficult to obtain for businesses in the non 'traditional' areas such as manufacturing than in the 'traditional' areas of food retailing and confectionery, tobacconists and newsagents (CTNs). Thus they point to the changing character of the Asian entrepreneur in the 1990s, as being more likely to be British born and educated to a high level, more likely to be seeking finance for a business outside the traditional areas and less likely to either wish to or be able to, rely upon finance from family sources. It is unclear whether the banks are yet targeting, or even aware of, this type of shift.

To summarize, there does appear to be clear evidence of market imperfections in the financing of black businesses in the United States. Fully comparable work has not been conducted in the United Kingdom but, that which has, does not suggest any clear market failure in the financing of Asian businesses. The only major reservation is whether banks are sufficiently aware of the 'new' type of entrepreneur emerging from the Asian community. So far as Afro-Caribbean businesses are concerned, these are clearly very different from Asian businesses and do experience significantly greater problems in both raising finance, and in their relationships with the banks. Whether this constitutes an illustration of a market failure, or whether it merely illustrates the higher risk associated with these types of businesses is not clear from the research conducted so far.

The assessment of risk is, of course, central to many of the arguments that point to market failures in the provision of bank finance to small firms. Risk assessment in turn hinges on information flows available to the lenders on the one hand and the borrowers on the other, with credit rationing and demands for collateral a banker's response to the information asymmetries that exist. A common response of bankers charged with creating debt gaps through credit rationing is that many of the proposals inhabiting the gap are not particularly well thought out or documented. Another is that high small firm failure rates reflect inadequate management planning and information flows, in particular relating to creditor and debtor positions and investment appraisal. The chapter by Nayak and Greenfield (Chapter 6) provides some salutary findings on this issue. Their focus is on the very smallest businesses, their management accounting information needs and practices. They provide data derived from a survey of 200 West Midlands micro businesses employing less than ten people, of which 123 were single person businesses with no employees. They show that of those businesses which had debtors, no fewer than 20 per cent kept no management accounting records of them, and a further 5 per cent could provide no information on what their debtors payment lag was. Furthermore in reply to the question, 'If you wish to purchase new equipment for your business please describe how you would decide if it was worthwhile' it appears that very few made reference to any kind of systematic investment appraisal. Even though it is to be expected that the smallest businesses will operate with relatively informal management structures, and in the case of sole proprietorships are principally putting their own capital at unlimited risk in their investment decisions, these findings are not encouraging. Moreover, as Nayak and Greenfield point out, businesses with the least systematic approach to these information issues are also, according to their proprietors, performing inadequately.

Nayak and Greenfield do not record the legal status of the members of their sample, but this may have an important bearing on the extent of information processed and recorded. As Nayak and Greenfield point out, there are a variety of external pressures on small businesses of all kinds to keep accounts and financial records. These include income and trading information for Inland Revenue and VAT purposes, and in the case of sole proprietorships and partnerships the accounting standards necessary to avoid charges of failing to keep proper books of accounts in the event of bankruptcy under the Insolvency Act of 1986. The pressures are however greatest for companies. In this case the reporting and disclosure requirements are often claimed to give greater credibility and status to incorporated as opposed to unincorporated businesses. The benefits of this, and of limited liability, as compared with the costs of meeting the accounting and disclosure requirements of the corporate form are the subject matter of the chapter by Freedman and Godwin (Chapter 7).

There was a boom in company incorporations in the 1980s. This was aided by streamlined registration procedures which meant, as the Institute of Directors approvingly noted, that 'an entrepreneur can acquire and begin to trade with a private limited company, ready made, in five minutes for a payment of £100' (IOD 1986). At the same time pressure to reduce administrative costs for the smaller companies led to important reductions in public disclosure requirements. By the end of 1992 a 'small company' (defined as one which fell below two of three size thresholds, viz. a balance sheet total of £1.4 million, employment of not more than fify, or turnover of £2.8 million) was required to submit only an abbreviated or 'modified' set of accounts for public scrutiny, consisting only of a brief balance sheet. In the face of all this Freedman and Godwin pose two fundamental questions. Should the privilege of limited liability be so easily accessible for small firms? (They estimate that 90 per cent of all companies meet the turnover definition of 'small' outlined above, and about one-third submit modified accounts.) Is the gain of limited liability and increased credibility associated with incorporation a genuine one for many of the small firms which have sought it even given the reporting dispensations noted above?

Freedman and Godwin provide answers based on the analysis of a mail questionnaire completed by 125 'small' limited companies, 80 per cent with a turnover of less than £1 million, and 146 unincorporated businesses. These businesses were asked about the reasons for their choice of legal form, their sources of finance and the costs and benefits of the legal form chosen. As might be expected about two-thirds of the company sample cited limited liability to bankers, creditors, and suppliers as a reason for incorporation, whilst half cited prestige and credibility. These were the two most frequently cited factors. However, over half of the respondents reported that their directors or spouses were currently providing personal guarantees, mainly to banks, in effect diluting the limited liability to which they attached such importance.

Perhaps the most significant finding of Freedman and Godwin is that whilst over 70 per cent of unincorporated businesses stated that owner's capital was important at business foundation, only 22 per cent of companies reported share capital as significant at incorporation. The authors conclude that 'The low response on use of share capital suggests that the formality of contributing share capital and the difficulty of retrieving it discourages contributions, and thus encourages the establishment of undercapitalized businesses.' They also note that where share capital was significant at incorporation the companies were larger at the time of the survey, had more directors and a more formal programme of directors' meetings. They were also more likely to be using other forms of finance such as debentures.

The upshot of their analysis is that for many small companies the costs of incorporation, in particular the statutory audit, are not worth the gains in terms of heavily diluted limited liability on the one hand, and supposedly enhanced prestige and credibility on the other, which may be of little use if future access to finance for growth is not a significant objective. In their view, the extent of capital commitment at incorporation is a more powerful signal of seriousness of intent than the act of incorporation alone. Although they doubt its political acceptability, Freedman and Godwin conclude that a substantial minimum capital requirement at incorporation would both serve to inhibit the smallest firms with least to gain from incorporating, and serve as a serious signal of commitment and financial viability for those that went ahead. The result may be fewer problems of failure due to abuse of the privilege of limited liability in undercapitalized firms, and the development in the longer run of more independent medium-sized businesses with a sounder financial structure.

The high rates of company formation of the 1980s were attended by another important development in the corporate sector, which affected the number of independent businesses surviving to maturity. After a decade of relative quiescence, acquisition and merger activity attained historically unprecedented levels (Hughes 1992). Although it is the largest of these transactions which capture the headlines, Cosh and Hughes, in the final chapter in this volume (Chapter 8) demonstrate the overwhelmingly small-scale nature of much acquisition activity in terms of the numbers of firms involved. Thus over 40 per cent of company acquisitions reported in the financial press in the late 1980s involved a consideration of less than £1 million, and between 30,000 and 40,000 VAT registered businesses, the vast majority even smaller than this, left the register as a result of acquisition by another business. Cosh and Hughes use the Cambridge Survey (SBRC 1992) to probe behind the motivations and characteristics of small firms involved as acquirers and targets in this merger wave and use post-merger company accounts to probe into merger effects. They show that in the period 1986–91 around one-fifth of their SBRC sample firms had made at least one acquisition, primarily in pursuit of scale gains, market share and diversified product lines. They also show, echoing the findings of Mason and Harrison, that acquisition is often regarded as an important exit route for potential sellers wishing to make capital gains, as well as helping to solve management succession problems. In the short-run, however, when incumbent management wishes to stay with the business, fears over loss of motivation, independence, and control leads amongst independent companies to a generally hostile attitude towards selling out in the short term. The analysis of completed acquisitions by Cosh and Hughes, using their company panel data set, suggests that in the SME sector the typical acquired company is a slower growing, less profitable business with lower liquidity and higher short-term debt than those which are not acquired. The acquirers on the other hand are larger and faster growing than non-acquirers, are not much more profitable than them, and do little initially to improve the profitability of the firms they acquire. This is especially so where the pre-merger performance of both firms was relatively good prior to merger. In these cases regression towards mean profitability may be an important part of the initially disappointing profits story.

THE IMPLICATIONS

To what extent does the evidence presented in these chapters support, or undermine, the view that access to finance has inhibited the contribution which small firms make to the UK economy? Do the chapters provide further evidence of 'gaps' or 'market failure'? If so, what should be done to rectify these matters?

The evidence in support of the existence of 'gaps' is provided mainly by Mason and Harrison in their analysis of 'business angels' (see Chapter 3). They suggest that many such individuals are seeking to make further investments in appropriate smaller companies, but are unable to obtain a sufficient flow of suitable proposals. Mason and Harrison also suggest there are business owners, whose firms are constrained in their growth by a 'shortage' of equity, and who would be prepared to share that equity with informal investors. The clear implication of their research is that information barriers exist between the two groups, and that it would be beneficial to the economy as a whole if these barriers were overcome.[5]

However, little evidence emerged from the other chapters that gaps exist in the financing of small firms, although it must be noted throughout that invariably such evidence arises from the analysis of *surviving* firms and severe financial constraints could have produced failure or takeover. This potential bias is met to some extent in Chapter 8 by Cosh and Hughes on takeover, whose results suggest that small acquisitions are likely to be characterized by relatively high debt levels and low profitability; a proportion, however, are not, and they would merit further analysis. Bearing this in mind, financial constraints did not appear in areas where they might be expected. For example, Jones, McEvoy and Barrett in Chapter 5 on the financing of white and ethnic businesses find no difference in this respect between Asian and white businesses. They do point to problems experienced by Afro-Caribbean business owners, but these may reflect both the sectoral composition and poorer performance of businesses established by these groups. The fact that they find that discrimination occurs does not, in itself, imply a market failure without further work on the relative characteristics of these businesses.

The same principles apply to the findings by Moore. He found that the businesses which are most constrained are those which are young and small – and therefore likely to appear most risky – and manufacturing businesses where the amount borrowed is likely to be higher than for service sector businesses. What Moore does not find, although he explicitly tests for it, is evidence that when age, size and sector are taken into account, businesses in the high technology sector are more likely to be finance constrained than those operating in the more conventional sectors. However, he does report that both groups claim financial constraints to be their most significant barrier to growth. What this points to is special care in dealing with young and small NTBFs and manufacturing firms. Distinguishing between those who want to grow and have the non-financial capacity to do so and those who don't, and dealing with the specific risk assessments of hi-tech firms, remain major challenges for lenders. The realities of the challenges are reflected in Chapter 6 by Nayak and Greenfield who show that, in some cases, lamentable weaknesses exist in financial decision taking in micro businesses.

The evidence presented in a number of the chapters is compatible with the view that the constraints in the development of the business lie as much with the entrepreneurs themselves, as with the financial institutions. A theme which emerges from the first chapter by Cosh and Hughes (Chapter 2) is the reluctance on the part of owners to share equity in the business with outsiders. Yet, as Keasey and Watson (1992) point out, this leads to the bank which provides short-term loans being involved in a contract in which they share the downside losses, but do not participate in the upside gains. A second aspect of this is the efforts by entrepreneurs to minimize their downside losses through transferring them to other parties, as reflected in the findings by Freedman and Godwin (Chapter 7), that directors of smaller companies are very much less likely to use their own capital in financing the businesses, than is the case for owners of unincorporated businesses. Here the evidence suggests that some entrepreneurs may (probably mistakenly) seek to use the credibility of incorporation as a substitute for their own resources in seeking loans from financial institutions.

So, what should be done? The evidence presented here provides little ammunition for the case of a widespread 'market failure' in the financing of small businesses. If there is to be a role for government it looks to be primarily that of encouraging the sharing of both the risks and returns of business ownership between the providers of finance and the entrepreneurs. One illustration of moving policy in this direction would be to specify, as Freedman and Godwin suggest, a substantial minimum share capital for those wishing to establish limited companies, leaving businesses with little aptitude for growth to remain unincorporated. Smaller businesses too may have a role to play here if together with, or independently from, banks they could form consortia to underwrite their own loan applications. This would capitalize on information held by the firms themselves and solve some of the information asymmetry problems which bedevil the lending market. The development of Mutual Guarantee Schemes along these lines is common in Europe. There is a seedcorn role for Government here in overcoming 'public good' problems in setting them up (see Hughes 1992).

The primary initiative for bringing about 'sharing' must lie, not with government, but with the parties themselves – the banks and the entrepreneurs. In principle, the task facing the banks can be easily specified, but is difficult to achieve in practice: it is to make better lending decisions. What this means is that, instead of seeking to maximize their market share as was the case in the late 1980s, banks seek to make better, if perhaps fewer, loans. One element of this may be to take on board the findings of Nayak and Greenfield which suggest that lending decisions favour those businesses where record keeping is better. Indeed the banks themselves may consider sponsoring training programmes for entrepreneurs in this area as being in their own interests.

The quid pro quo is that banks have to shift emphasis away from collateral-backed short-term lending – towards more frequent equity participation or at least towards longer-term lending. Cosh and Hughes note that the latter has happened to some degree but it needs to go further. In this way the banks can build a longer-term and more fruitful relationship with their client base. They are also likely to develop the information framework which brings together informal investors and entrepreneurs as recommended by Mason and Harrison. In a more trusting relationship their own initiatives to provide small sums of venture capital are also more likely to be successful.

Concl

The history of the provision of small sums of equity or seed capital has not been filled with success. In essence this is because the entrepreneur, who is totally convinced that the business proposition will be hugely successful, is very unlikely to be willing to share the fruits of that success. The proposals therefore which are put to venture capital organizations or individuals are likely to reflect a high proportion of cases where the entrepreneur is simply seeking to reduce his or her own financial input into the venture – i.e. minimize their downside risk.

It is then in the overall interests of the small business community and the UK economy as a whole that a greater sense of trust is encouraged between the entrepreneurs and the financial institutions. The role of government has to be to ensure an even-handed treatment of both sides, with a particular focus on educating entrepreneurs and bankers that the contract which provides finance from outside has fully reflected both the upside gain as well as the downside losses. That requires both sides recognizing the benefits of a longer-term relationship than currently exists.

NOTES

1 We thank all contributors to this volume for their comments on our Introduction. The views expressed are ours alone and not necessarily those of the contributors nor the sponsors of this research.
2 Some businesses angels could be classified as partners/working shareholders, so that a wider definition of 'informal funds' might increase its apparent importance somewhat.
3 This amount comprises both investments in the previous three years plus amounts available for investment.
4 Some TECs have already operationalized this 'match making' function.
5 Mason and Harrison also imply that latent demand for equity finance would be realized if angels were seen to be more accessible. They regard the key step to be a demonstration to business owners that such angels exist and are prepared to make small investments.

REFERENCES

Advisory Council on Science and Technology (ACOST) (1990) *The Enterprise Challenge: Overcoming Barriers to Growth in Small Firms*, HMSO, London.

Bank of England (1993) 'Bank lending to smaller businesses', The Bank of England, London.

Barber, J., Porteous, M. and Metcalfe, S.J. (eds) (1989) *Barriers to Growth in Small Firms*, Croom Helm, London.

Bates, T (1991), 'Commercial bank finance in white and black owned small business start ups', *Quarterly Review of Economics and Business*, 31, 3, Spring, 64–89.

Binks, M.R., Ennew, C.T. and Reed, G.V. (1992a) *Small Businesses and their Banks: 1992*, Forum of Private Business, Knutsworth Cheshire.

Binks, M.R. Ennew, C.T. and Reed, G.V. (1992b) 'Information asymmetries and the provision of finance to small firms', *International Small Business Journal*, 11, 1, 35–46.

Cowling, M., Samuels, J. and Sugden, R. (1991) *Small Firms and the Clearing Banks: A Sterile, Uncommunicative and Unimaginative Relationship*, Association of British Chambers of Commerce, London.

Deakins, D., Hussain, G. and Ram, M. (1992) *Finance of Ethnic Minority Small Businesses*, University of Central England in Birmingham.

Harrison, R.T. and Mason, C.M. (1992), 'The roles of investors in entrepreneurial companies: a comparison of informal investors and venture capitalists', Venture Finance Research Project, Working Paper no. 5, University of Southampton (Urban Policy Research Unit, Department of Geography) and University of Ulster (Ulster Business School).

HMSO (1971) *Report of the Committee of Inquiry on Small Firms*, (Bolton Report) Cmnd 4811, HMSO, London.

HMSO (1979) *Interim Report on the Financing of Small Firms*, Cmnd. 7503, HMSO, London.

HMSO (1991) *Constraints on the Growth of Small Firms*, HMSO, London.

Hughes, A. (1992) 'The problems of finance for smaller businesses' University of Cambridge, SBRC Working Paper no. 15 in Dimsdale, N. and Prevezer, M (eds) *Capital Markets and Company Success*, Oxford University Press, Oxford, 1993.

Institute of Directors (1986) *Deregulation for Small Private Companies*, IOD, London.

Jones, T., McEvoy, D. and Barrett, G. (1994) 'Raising capital for the Ethnic Minority Small Firm' in Hughes, A. and Storey, D.J. (eds) *Finance and the Small Firm*, Routledge, London.

Keasey, K. and Watson, R. (1992) *Investment and Financing Decisions in the Performance of Small Firms*, National Westminster Bank, London.

Monck, C.S.P., Porter, R.B., Quintas, P.R., Storey, D.J. and Wynarczyk, P. (1988) *Science Parks and the Growth of High Technology Firms*, Croom Helm, London.

Myers, S.C. and Majluf, N.S. (1984) 'Corporate financing and investment decisions when firms have information that investors do not have', *Journal of Financial Economics*, 13, 187–221.

Oakey, R.P. (1984) *High Technology Small Firms*, Frances Pinter, London.

Pavitt, K., Robson, M. and Townsend, J. (1987) 'The size distribution of innovating firms in the UK: 1945–1983', *Journal of Industrial Economics*, 45, 296–306.

Roberts, E.B. (1991) *Entrepreneurs in High Technology: Lessons from MIT and Beyond*, Oxford University Press, New York.

SBRC (1992) *The State of British Enterprise: Growth, Innovation and Competitive Advantage in Small and Medium-sized Firms*, Small Business Research Centre, Cambridge.

Ward, R (1991) 'Economic development and ethnic business' in Curran, J. and Blackburn, R. (eds) *Paths to Enterprise: The Future of Small Businesses*, London, Routledge.

Ward, R. and Reeves, F. (1980) *West Indians in Business in Britain*, HMSO, London.

Wetzel, W.E. Jr. (1987) 'The informal venture capital market: aspects of scale and market efficiency', *Journal of Business Venturing*, vol. 2, 299–313.

Wilson, P. and Stanworth, J. (1987) 'The social and economic factors in the development of small black minority firms: Asian and Afro-Caribbean businesses in Brent, 1982 and 1984 compared' in O'Neill, K., Bhambri, R., Faulkner, T. and Cannon, T. (eds) *Small Business Development: Some Current Issues*, Avebury, Aldershot.

2 Size, financial structure and profitability: UK companies in the 1980s

Andy Cosh and Alan Hughes[1]

INTRODUCTION

This chapter discusses the financial structure and profitability of UK businesses in the period 1977–91. It considers the ways in which size is related to the balance sheet structure, financing and profitability of these businesses. Unlike most studies of business finance which focus on smaller concerns we do so in a way which allows us to make direct comparisons with large companies. Thus as well as making use of a specially conducted national survey of over 2,000 businesses employing less than 500 people in 1991, which *inter alia* provides share ownership financial and other qualitative data for the years 1987–90 (SBRC 1992) we also make use of a specially constructed panel data set of a size stratified sample of several thousand companies covering the entire corporate sector in the period 1977–83, using information derived from their accounts lodged at Companies House (Penneck 1978, Lewis 1979, Erritt 1979). These two data sources allow us to make size comparisons both *within* the SME sector, and between that sector and the remainder of the business population. Although our findings bear on the existence or absence of 'debt gaps' or 'equity gaps' facing small UK firms our main concern is not to resolve that debate but to shed light on what the relative financial and performance characteristics of smaller firms compared to larger ones actually are. In doing so we refer to changes since the last major surveys of small firm financing in 1971 and 1979 (HMSO 1971, HMSO 1979). We also make use of developments in the theory of finance since those reports to shed light on the question of how financial structure and performance might vary with firm size, why in principle equity and debt gaps might arise and what guide, if any, variations in patterns of financial structure across companies of different sizes provide to the existence of such gaps. The majority of our data relates to companies and the chapter begins, in the next section, by placing this form of business organization in context with other legal forms in the UK, and by then examining the way in which share ownership varies with company size.

We then consider why different forms of organization exist and the implication for the financing and growth of enterprises which are either sole proprietorships, partnerships or closely held companies with negligible or zero outside equity interests. This discussion is followed by a section which briefly reviews the ways in which the theory of corporate finance bears on the determination of financial structure in smaller firms in the light of their ownership and risk characteristics compared to larger firms. This is followed by the substantial empirical section of the chapter which analyses variations in financial structure and profitability by firm size and also examines survey evidence on perceived financial constraints on growth, and sources of new finance again allowing for variations due to size. A final section summarizes our principal conclusions.

BUSINESS ORGANIZATION, OWNERSHIP AND RISK IN THE SME SECTOR

The mainstream theoretical literature on company finance deals with the financing decisions of companies with large numbers of outside stockholders, and with a choice between raising finance externally either via debt, or via further equity issues to new or existing shareholders. In general the decision takers are assumed to be salaried professional managers. These either act voluntarily in the shareholders' interests, or are persuaded into doing so by carefully designed incentive packages, or coerced into doing so by a variety of product market, capital market, or managerial labour market pressures. This literature has focused on particular problems when owners and managers are functionally separated and costs arise for owners in their attempts to ensure that businesses are managed in the owner's interests. Thus the transactions cost and agency models of the firm draw links between the structure of finance and its implications for 'bonding' the managers and owners together; the transactions costs of monitoring behaviour; and the design of incentive payments systems linking managers pay to the performance of profits or equity stock. (Williamson 1985, Jensen and Meckling 1976, Fama and Jensen 1985). It has also focused on the implication of risk and financial distress both for the shifting of its costs between equity and debt holders, and for capital structure. In particular since financial distress is more likely for firms with high business risk this literature suggests they should, in general, issue less debt relative to equity.

Recently the continuing policy interest in non-quoted and very small firms has led the very large empirical literature on their financial structure and problems to be supplemented by attempts to explain it in terms of recent theoretical developments concerning listed stock market companies (Ang 1991). In doing so an important starting point is to recognize and document the fundamental differences in risk characteristics and share ownership characteristics, which separate the quoted and non-quoted sectors, and to place small unquoted companies themselves in the wider context of different business forms such as partnerships.

The numbers and size distribution of businesses by legal type in the UK is shown in Table 2.1. It includes those businesses which met the minimum VAT thresholds in 1992. There were 1.66 million of them and just over half a million were companies. Of these however only around 2,500 were sufficiently large and with a sufficiently dispersed stock ownership to be listed on either the UK International Stock Exchange or the Unlisted Securities Market (USM) (Pawley, Winstone and Bentley 1991). Companies are massively outnumbered by partnerships and sole proprietorships, which accounted for nearly two-thirds of VAT registered businesses in 1992. The table shows, however, that companies dominate as the favoured business form when we consider the larger size classes by turnover. Thus for all industries taken together we find that 76 per cent of sole proprietors had a turnover of less than £100,000 compared to 53 per cent of partnerships and only 40 per cent of companies. Conversely only 0.6 per cent of proprietorships had a turnover greater than £1 million whilst 2.3 per cent of partnerships and 18.2 per cent of companies did. The table shows that companies are relatively unimportant in the more labour intensive small-scale construction and service industries and more significant in manufacturing where they account for 58 per cent of all registered businesses.

How does risk of business failure vary across these size classes? It is apparent that as far as firms registered for VAT are concerned the mortality rate amongst any cohort of newly formed, and hence small firms, is very high. Thus 50 per cent will have deregistered after five years and 75 per cent after twelve years (British Business 1987). As a reflection of this turbulence we find that between 1980 and 1990 the growth of businesses registered for VAT of 420,000 was the difference between 1.7 million deregistrations and 2.1 million registrations. In the company sector alone it is also clear that smaller businesses fail more frequently with dissolution amongst those with less than £1m of total assets almost three times as high as in larger companies (Cosh and Hughes, this volume, Chapter 8). Other things being equal this higher degree of riskiness should predispose smaller firms to a low debt to equity ratio. Other things may not, however, be equal and the equity position in small firms must now be considered.

It is clear that sole proprietorships and partnerships have no 'outside' equity. The unimportance of outside equity is often also taken as a definition of 'smallness' within the company sector (HMSO 1971). For young small businesses this is often a reflection of their relative youth and the extent to which start up is closely connected with internal financing from personal and family sources, a feature which is of long standing significance in the UK (Shaw 1993). There is, however, a relative paucity of precise data on how share ownership does actually vary by company size beyond start up. Table 2.2 goes some way towards addressing this question. It shows the results of analysing the share ownership characteristics of UK businesses employing less than 500 people responding to the Cambridge University National Small and Medium Sized Firms Survey in 1991 (SBRC 1992). Of these 45 per cent had a turnover of less than £1,000,000 in that year. The median percentage of shares held by the boards of the companies responding as a whole was 100 per cent. In fact 75 per cent of all boards held over 75 per cent of the stock. Amongst micro and small businesses employing less than 100 people over 80 per cent of the boards held 100 per cent. Only for companies employing between 200 and 500 workers did the median holding fall, and then only to 75 per cent and nearly half of the boards continued to hold over three-quarters of the stock. The table also shows that younger firms were more likely to be closely held than older firms, although in both cases whether the firm was born before or after 1980 the median holding remained at 100 per cent. Thus we might expect companies with turnover of less than £1,000,000 to be predominantly closely held.

Once businesses grow beyond the 500 employee scale represented by the Cambridge sample, and especially if they become quoted, we might expect more share dispersion to occur. Even here, however, substantial board share ownership persists. Thus a study of all the quoted companies in the UK food, electrical engineering, and bricks, pottery and glass industries in 1980 revealed that in 20 per cent of cases the board held over 50 per cent of the stock (Cosh and Hughes

Table 2.1 The distribution of businesses in the UK registered for VAT by size, sector and legal form in 1992

Sector	Distribution of businesses by legal type %	Total number of businesses (000s)	Distribution of businesses by turnover size band (£000s)				
			1–99 %	100–499 %	500–999 %	1,000 + %	All %
Agriculture, Forestry, Fishing[1]							
All businesses	100.0	164	72.3	24.3	2.2	1.1	100
Sole proprietorships	44.9	73	86.5	12.6	0.7	0.2	100
Partnerships	48.7	80	64.2	32.7	2.4	0.8	100
Companies and public corporations	6.0	10	32.3	44.3	12.5	10.8	100
Production							
All businesses	100.0	162	42.9	32.9	9.1	15.1	100
Sole proprietorships	24.4	40	76.6	21.9	1.1	0.3	100
Partnerships	17.2	28	54.3	40.1	4.1	1.5	100
Companies and public corporations	58.2	94	25.3	35.5	13.9	25.4	100
Construction							
All businesses	100.0	241	62.2	27.0	5.2	5.5	100
Sole proprietorships	51.2	123	81.7	17.1	0.9	0.2	100
Partnerships	18.4	44	59.0	35.7	3.9	1.5	100
Companies and public corporations	30.4	73	31.5	38.4	13.4	16.7	100

Transport and other services

All businesses	100.0	1099	55.6	31.6	5.6	7.1	100
Sole proprietorships	40.1	440	72.7	24.8	1.7	0.7	100
Partnerships	24.6	270	48.6	43.2	5.2	3.0	100
Companies and public corporations	32.8	360	40.2	30.8	10.9	18.2	100
General government and non-profit bodies	2.5	29	52.8	38.2	2.8	6.3	100
All Industries							
All businesses	100.0	1664	57.0	30.4	5.5	7.1	100
Sole proprietorships	40.6	676	76.1	21.9	1.4	0.6	100
Partnerships	25.4	422	53.0	40.2	4.4	2.3	100
Companies and public corporations	32.3	537	40.2	30.8	10.9	18.2	100
General government and non-profit bodies	1.7	29	53.0	37.7	2.9	6.4	100

Source: Business Monitor PA 1003 *Size Analysis of UK Business 1992*, HMSO, London

Note: 1 All businesses includes general government and non-profit making bodies. The latter are however of significance only in the transport and other services sector. A separate analysis for them is therefore shown only for that sector and for all sectors taken together. For the other sectors it follows that the sum of partnerships, proprietorships and companies and public corporations is somewhat less than the total for all businesses shown in the respective sector.

Table 2.2 The distribution of businesses by the percentage of shares held by the board in a sample of UK SMEs 1991

% Shares held by board	All %	Older[1] %	Newer %	Micro %	Small %	Medium %	Larger %
0	4.8	6.1	3.7	2.3	3.4	7.2	18.5
1–24	2.9	3.4	2.4	1.8	1.9	5.9	8.9
25–49	5.8	7.5	4.2	2.8	5.7	9.2	10.3
50–74	11.3	11.9	10.8	11.8	11.0	11.8	13.0
75–99	15.6	17.1	14.6	14.1	15.3	20.3	16.4
100	59.6	53.9	64.4	67.2	62.8	45.8	32.9
Median board holding	100.0	100.0	100.0	100.0	100.0	93.0	74.0
Sample size	1614	799	789	390	911	153	146

Source: SBRC (1992)
Notes: 1 Older formed pre-1980; Newer formed 1980 or later; Micro 1 < 10 employees; Small 10 < 100; Medium 100 < 200; Larger 200 < 500.

Table 2.3 Median board holdings and the median value of the sum of non-board holdings over 5 % in the 600 largest non-financial UK companies in terms of market value in 1982

Company rank by market value	Median board holding %	Number of, and median value of the sum of non board holdings over 5%					
		All		Financial institutions		Other	
		No.	Median %	No.	Median %	No.	Median %
1–100	0.2	29	8.7	16	6.6	13	23.0
101–200	0.6	71	11.5	56	6.6	15	9.6
201–400	11.6	206	14.8	147	11.2	59	17.8
401–600	19.5	267	19.3	198	11.4	69	14.8

Source: *Crawford's Directory of City Connections 1982/3 Stock Exchange Companies*, March 1982

1989). Moreover, Table 2.3 shows that even amongst the very largest UK quoted companies board members hold between them quite significant blocks of stock. Companies ranked between 200 and 400, in terms of UK stock market value, have median boards holding of over 10 per cent and those ranked between 401 and 600 have median holdings of nearly 20 per cent. The distinction between these companies and even the largest SMEs is however clear cut. Table 2.3 also shows that for the largest quoted companies the principal non-board shareholdings are not in the hands of individuals but of financial institutions such as pension funds and insurance companies. For instance for companies ranked 1–100 there were in total twenty-nine non-board holdings each of over 5 per cent. The median sum of these holdings in any one company was 8.7 per cent. Of the twenty-nine, sixteen were held by financial institutions and the median value of the sum of these holdings in a company was 6.6 per cent.

These institutions as a group are the largest shareholders in the largest companies. Tables 2.4 and 2.5 show that for SMEs the shareholdings of financial (and non-financial businesses) are relatively insignificant except for firms employing between 200 and 500 employees, and that the largest shareholder is overwhelmingly likely to be a single individual (holding as Table 2.4 shows a median value of 50 per cent of the stock). The absence of significant institutional holdings is unsurprising given the transaction costs of dealing with a small holding in value terms and the thin market in shares the majority of which remain closely personally held (Cosh, Hughes and Singh 1990). Thus in the UK it is reasonable to describe firms employing less than 500 employees as closely held, and as characterized by dominant personal shareholdings. The smaller and younger the firm the stronger this characterization becomes.

It is important to note that in addition to share-owning managers who are the directors of their companies, most of these closely held businesses will also employ non-board salaried managers. Thus the Cambridge Survey reveals that over 90 per cent of SMEs employing more than ten people had at least one managerial post in the labour force, and 60 per cent of those employing less than ten did (SBRC 1992: 50). These, however, are clearly in a different control position to the non-share owning directors of managerial models of the firm (see also Storey 1990).

In terms of the agency models of firm organization the existence of tightly held companies, and the large numbers of proprietorships and small partnerships represents an optimal trade off between the gains of low agency, monitoring and bonding costs (ownership and management are fused, with owners being the residual claimants on the outcome of their own self-monitored efforts); and the losses of restricting investments, and the scale of activity, to the limits imposed by the human and financial capital of the owner-managers, and by self-financing through retentions. The majority of new business starts are financed by personal and family savings, and loans raised on personal assets (e.g. second mortgages on houses) (Mason 1989, Reid and Jacobsen 1988, Turok and Richardson 1991). Subsequent to start up firms may choose to remain small and eschew growth which would

Table 2.4 The distribution of UK SMEs in 1991 by the percentage of shares held by largest shareholder

% Shares held by largest shareholder	All %	Older %	Newer %	Micro %	Small %	Medium %	Larger %
1–24	6.2	7.5	5.0	2.1	5.8	9.4	17.5
25–49	27.4	27.9	27.1	19.2	29.8	34.9	27.3
50–74	38.8	38.2	39.1	42.1	39.9	32.2	29.4
75–99	20.4	19.4	21.2	27.9	18.9	17.4	13.2
100	7.1	6.9	7.6	8.7	5.9	6.0	12.6
Median size of largest holding	50.0	50.0	51.0	50.0	50.0	50.0	50.0
Sample size	1,600	795	778	390	904	149	143

Source: SBRC (1992)
Notes: See Table 2.2.

Table 2.5 The distribution of UK SMEs in 1991 by type of largest shareholding

Type of largest shareholding	All %	Older %	Newer %	Micro %	Small %	Medium %	Larger %
Individual	88.9	88.5	88.9	92.8	91.4	82.8	68.0
Non-financial business	3.2	3.0	3.6	1.5	3.0	4.6	8.7
Financial business	3.8	3.5	4.3	2.2	3.1	7.9	9.3
Other	4.0	5.0	3.1	3.5	2.5	4.6	14.0
Sample size	1,669	833	805	403	947	152	150

Source: SBRC (1992)
Notes: See Table 2.2.

involve accepting either ownership dilution through issuing equity, or restrictions on freedom of action such as covenants or financial reporting requirements which may accompany the use of anything other than short term debt (Norton 1991, Barton and Matthews 1989). That owner-managers place great store by independence is clear from the abundant survey data on their motivations for setting up and running businesses. The emphasis on 'running your own show' and 'working for yourself' were noted in the Bolton Report as 'psychological satisfactions which appeared to be much more powerful motivations than money or the possibility of large financial gains' (HMSO 1971: 23).

This generalization has been confirmed by numerous studies (Collins and Moore 1979, Kets de Vries 1977, Cooper 1986, Caird 1989, Hirisch 1986, Goffee and Scase 1985). It is also reflected in responses in the Cambridge Survey to questions about the disadvantage of being taken over. Loss of independence and control massively outscored any other reason (SBRC 1991: 47 and Cosh and Hughes, this volume, Chapter 8). It is, however, a generalization and not all closely held companies may be so by choice, or intend to remain so. The characteristics of original owners or founders in terms of human capital, risk aversion, and personality may have a considerable impact on their objectives (Bragard *et al.* 1985, Chell 1985, Chell *et al.* 1991, Collins and Moore 1979, Deeks 1972, Lafuente and Salas 1989, Stanworth and Curran 1973, Nicholson and West 1988, Watkins 1983 and Storey 1982). The small business literature is replete with typologies of small business owners which contrast for example craftsmen owners with promoter/opportunists on the one hand and manager/trustees on the other. Of these only the latter two may be interested in profitability or expansion as primary objectives with the 'promoter' emphasizing short-run performance to be followed by exit through the sale of the business, and the 'manager' more interested in the 'long run growth of the business and the construction of an administrative hierarchy' (Hornaday 1990). As with all typologies this may be criticized and there are several studies exploring variations in owner-manager type (e.g. Stanworth and Gray 1991 Ch. 7, Chell *et al.* 1991). The significant point which emerges, however, is that a very small proportion of companies may be owned and managed by individuals concerned with rapid long-run growth (Vickery 1989). Thus in a survey of businesses employing less than fifty people Hakim (1989) puts the proportion planning rapid expansion into new products or markets as low as 10 per cent, with a further 35 per cent aspiring to 'slow steady growth'. The Cambridge National Survey reported substantial planned growth for 17 per cent of firms employing less than ten workers and 22 per cent of firms employing between eleven and forty-nine workers (SBRC 1992: 29) in a survey which may, if anything, be biased towards faster growing firms who have sought credit ratings on the Dun and Bradstreet database. Moreover it is also known that actual growth rates of small firms are highly skewed with only a handful of any cohort accounting for the bulk of expansion in employment terms in any period (Storey and Johnson 1987).

A final distinction is that between innovative and established firms. This has a bearing both upon riskiness and capital requirements.

Established SMEs in mature niche markets will have better track records and lower investment requirements than innovative firms seeking to break into new territory. The latter might be faced by a high risk profile in terms of their project area and lack an established track record both of which may affect the form, stability and cost of finance.

Finally it is important to recognize that compared to large quoted companies smaller firms, especially in the early stages of growth, are faced with relatively discontinuous financial requirements so that their demand for funds may be lumpy and often connected to other major step changes in their management organization and geographical and product market spread (ACOST 1990, Bannock *et al.* 1991). In this sense small firms which do expand may be seen as progressing through a series of financing stages culminating in stable maturity as an independent SME, sell out or listing on a stock exchange. A stylized picture of these stages for the UK including rescue and replacement capital to supplement that of the original owners is shown in Table 2.6 along with an indication of the scale of finance typically involved and its riskiness. We should therefore expect to find considerable variation and skewness around mean values of financial ratios for smaller firms.

To conclude smaller unquoted companies differ from their larger quoted counterparts in terms of age, riskiness, and the closeness with

Table 2.6 Financing stages

	Typical size	Timescale (years)	Risk
Early stage financings			
Seed capital	£5k–£100k	7–12	Extreme
Start up	£100k–£300k	5–10	Very high
Second-round	£300k–£1m	3–7	High
Later stage financings			
Mezzanine/development			
bridge	£500k–£2m	1–3	Medium
Last round before			
planned exit	£1m+	1–3	Low
Buy-outs/buy-ins	£1m–£25m	1–3	Low/high
All stages			
Turnaround			
New management			
and rescue finance	£100k+	3–5	Med/high
Replacement capital	£5k–£5m	1–3	Low

Source: Bannock *et al.* 1991

which they are held. Moreover within the small business sector itself we may expect considerable variations in financial requirements and attitudes to risk and independence depending upon their size, age and innovation characteristics. Thus in the small firm as in the larger firm sector management strategy may play a key role in the determination of capital structure (Friend and Lang 1988, Pettit and Singer 1985, Norton 1990). In the discussion of theory and evidence which follows we will focus on variations in financial structure and performance by size. The particular characteristics of innovative or hi-tech firms are analysed using the Cambridge Survey data and other sources in Moore (this volume, Chapter 4). An analysis of the impact of age is outside the scope of this chapter, and is left for another occasion.

OPTIMAL FINANCING AND THE SMALL FIRM

In this section we consider what light, if any, may be thrown upon patterns of small business finance by the existing theories of optimal capital structure derived principally from the analysis of large corporations. In doing so we will assess the implications of allowing for the ownership and risk characteristics of smaller firms discussed in the previous section.

The capital structure puzzle which has exercised finance theorists is concerned with the question of whether or not there exists an 'optimal' capital structure which contains both debt and equity. A number of possible arguments have been put forward to explain the existence of such an optimal structure emphasizing respectively the impact of tax breaks versus bankruptcy costs in the determination of upper limits to borrowing (Stiglitz 1972, Altman 1984); the relative agency costs of raising debt versus equity (Jensen and Meckling 1976); and the impact in imperfectly informed markets of using capital structure as a signalling device (Ross 1977, Myers 1984, Myers and Majluf 1984). Of these the last has led to the development of the Pecking Order Hypothesis (POH) of financial structure in which the financing of projects is undertaken by first using internal resources then debt and, as a final resort, equity. The essential insight of this hypothesis applied to public listed companies is that, faced with investment projects yielding positive net present values, the managers of such companies will put the interests of existing shareholders first. If managers have information about the likelihood of success of projects which potential investors in equity do not have, then the latter will take the issuing of equity as a signal that the existing stock is overvalued. This is because they reason that if an investment project was thought by the managers to lead to an increase in market value they would prefer to preserve this for existing shareholders and not dilute their gain. Thus new investors end up not being able to distinguish between the 'good' news (a positive net present value project) and the 'bad' news (equity issues implying overvaluation). If new equity were the only source of funding some positive net present value projects would not be undertaken. To combat this problem listed companies can use retentions to build up excess liquid assets to fund investment opportunities as they occur. If that is insufficient then debt will be preferred to equity because its payouts are less correlated with expected future payoffs, are therefore less risky and do not carry the adverse signalling implications of equity. Equity will follow when firms are recognized as fully leveraged. The POH is dynamic and implies that individual firms may have very different debt to equity ratios depending upon the matching of their profitability with their investment opportunities. Low-profit high-investment opportunity firms will tend to have higher debt and ultimately issue more equity than high-profit low-investment opportunity firms.

This hypothesis can be readily applied to smaller unquoted firms. First their closely held nature makes more natural the Myers/Maljuf hypothesis that managers will act in the interest of 'old' shareholders. The managers themselves are all usually the 'old' shareholders. Second, the information asymmetry which gives rise to the problem in the first place may be greater for small firms especially those which are breaking new ground, or are recently founded. Moreover the likelihood of these information gaps being plugged is much less for small unquoted firms since the market for information about them is restricted. Few potential equity investors will find it worthwhile to pay for information about companies whose stocks are closely held and may at best be very thinly traded and illiquid when they do come on the market (Holström and Tirole 1990). Thus in terms of the POH, the desire for internal funding to meet investment opportunities and the need to keep excess liquid assets to meet the relatively discontinuous 'lumpy' transitions which smaller firms face will be reinforced for them compared to large firms.

In these circumstances faced with a shortfall of funding from internal resources to meet investment plans, reliance will be placed on debt. Here too the risk characteristics of small firms as a class and the relative lack of information about the quality of their owner-managers and their projects may influence their debt position relative to large quoted firms. Once again information asymmetries lie at the root of the problem and suggest that new and small firms may be at a disadvantage compared to large firms. In the debt story borrowers know more about the expected riskiness of their project than lenders, especially if they are small or new. There is likely to be less public information about their past performance on which lenders may base their decisions. If lenders try to use interest rate increases to allow for their perceptions of riskiness they are likely to drive out those who believe on their own private information that they are less risky than the bank does, and to attract those who believe they are more risky. The result for the bank will be a higher risk profile for its loans than it bargained for. Given two borrowers with the same expected mean return but different spreads around the mean, it is the borrower who believes he has the wider spread who is more likely to take up the loan. In that sense the good loans drive out the bad. The response to this may be to avoid interest rate increases as a discriminatory tool and substitute credit rationing by refusal to supply loans, driving out some socially desirable investment in the process (Stiglitz and Weiss 1981). This argument is sensitive to the assumption that projects differ only in the spread around mean values. Consider two borrowers each of whom will earn the same return if their project succeeds but holding different views about the likelihood of success (and hence about their expected mean return) and a lender happy to share their view about the return if the projects succeed but uncertain about which borrower is more likely to succeed. The problem facing the lender is to charge an interest rate which yields the lenders required return, given the lenders views of the likelihood of each of the borrowers failing. Faced with an interest rate premium to reflect the lenders perception of the risk of failure of the borrowers taken together it is now the borrower who believes he has a lower risk of failure who will be most attracted. Those whose private information leads them to believe that they have a higher chance of failure and whose expected mean return is therefore lower will at the margin withdraw (DeMeza and Webb 1987). In practice there is likely to be uncertainty about both expected means and expected riskiness, and therefore some uncertainty about the usefulness of interest rates as a screening device.

An alternative to using interest rates to solve the risk problem is to seek collateral. This helps elicit information about the borrowers risk perception. It also reduces the banks downside risk in the event of a lender obtaining a loan and then pursuing higher risk/higher return projects than the one for which it was granted (the moral hazard problem) (Bester 1987). The collateral route has disadvantages. It erodes the limited liability of the owner. It also raises problems for those with positive net present value projects, but who are not already blessed with assets to back their judgement. Moreover the more conservative the valuation of the collateral the higher the disincentives for these firms. Recent evidence for the UK suggests that there is relatively little effort to appraise individual project risk and use the interest rate to price for risk differentials. The more usual route is the pursuit of collateral and secured loans and the use of conservative 'carcase' valuations of collateral (HMSO 1991, Binks *et al.* 1988). In a sample of over 3,000 UK firms in 1990 only 18 per cent had not been required to put up collateral for overdrafts or loans (the vast majority of these cases involved overdraft facilities of less than £20,000 where the administrative costs make the taking up of collateral not worthwhile). Where collateral was taken on overdrafts the ratio of collateral to overdraft limit was on average 4.4 and of collateral to overdraft use was 9.2. For loan facilities the ratio was significantly higher for the smallest firms (i.e. with turnover of less than £500,000) (Binks, Ennew and Reid 1990). The smaller size classes contain a higher proportion of younger start up firms who may therefore face the asset constraint problem in a severe form. It is widely agreed that these companies may be forced as a result to be more reliant on retentions and personal funds than more mature businesses who at least have a 'carcase' to value (Churchill and Lewis 1983, Walker 1989, Evans and Jovanovic 1989). They may also seek to obtain assets via hire purchase or leasing rather than purchasing them using borrowed funds. The reclaiming of these assets by the leasing company in the event of failure being a much more straightforward affair than the resolution of competing creditor claims over business owned assets in bankruptcy.

Given these potential problems in the debt market (which reflect the relative riskiness of small firms as a group, and the relative lack of information to assess individual risk) a further refinement of the POH may be advanced for smaller firms. If as we have argued a significant proportion are concerned primarily with the maintenance of independence and freedom from control then this will also influence the form of debt sought. In particular credit which comes with the least formal restrictions will be favoured, such as leasing, hire purchase, trade credit and short-term loans such as overdrafts. The latter in particular combine flexibility with an absence of the kind of regular monitoring and repayment of interest that go with fixed term and longer loans and which have led many to argue that the latter are an optimal method of solving some of the lenders agency and moral hazard problems (Aghion and Bolton 1992, Moore 1992, Hart and Moore 1990). The trade off for the users of these sources of funds is that whilst both benefit from informal trust relationships, both as a consequence may be subject to curtailment or pressure without the protection of formal legal agreements which longer-term debt contracts usually possess. Thus customer and bank pressures in recessions when all parties are faced with adverse market circumstances may lead to severe problems for smaller firms who choose to rely extensively on these forms of finance. It is no coincidence that complaints about bank–small firm relationships are distinctly anti-cyclical, and are all the more intense because of the perceived market power of large banks, and of large customers putting pressure on trade debts.

To sum up our discussion. The particular features of small firms would lead us to argue that within an overall POH framework we would expect that compared to larger firms with more dispersed equity smaller closely held firms would

1 rely more on carrying 'excess' liquid assets to meet discontinuities in investment programmes;
2 rely more on short-term debt including trade credit and overdrafts;
3 rely less on new shareholders equity compared to 'internal' equity and to debt in raising new finance; and
4 rely to a greater extent on hire purchase and leasing arrangements.

We would, however, expect there to be a very wide variation in these patterns of finance across small firms, because of differences in the life cycle position of firms, their size and their strategies towards independence and growth.

We address these questions in the next section, which in addition to discussing patterns of financial structure by company size also examines relative profitability.

THE COMPARATIVE FINANCIAL STRUCTURE AND PROFITABILITY OF LARGE AND SMALL COMPANIES

Table 2.7 presents a summary analysis of the balance sheet structure, gearing, and profitability of large and small UK non-financial companies in the UK in the period 1987–9. 'Large' companies are those ranked in the top 2,000 in terms of capital employed in the UK non-financial corporate sector. They are primarily quoted companies. 'Small' companies are a one in 300 sample of the remainder of the corporate sector stratified by size of capital employed. Averages for these two groups may obviously conceal wide variations within them, and may also reflect the effects of aggregation across industries characterized to different degrees by the presence of large and small firms. As a rough check on the latter effect a breakdown of the sample is provided into manufacturing and non-manufacturing industries. The data are the best available official statistics for the purpose in hand and may be compared with similar analyses specially carried out in the Bolton and Wilson Reports dealing, *inter alia*, with small business finance (HMSO 1971, 1979). It is worth emphasizing, however, that in the course of the 1980s increasing numbers of small companies took advantage of the dispensation to submit modified accounts in their returns to Companies House. This has led to an increase in the estimation involved in producing the data which is reported here. It is ironical that at a time when information on small company performance is of growing interest it is becoming increasingly difficult to obtain it in a useful form. Consequently when we wish to probe beyond these aggregate figures we use our own specially constructed panel sample for an earlier period 1977–83 (which predates the extensive use of modified accounts) and our Cambridge National Survey Results for 1987–90.

The data in Table 2.7 are presented as averages over the three financial years 1987–9. Each balance sheet item is shown as a percentage of total assets/liabilities whilst rates of return are shown upon calculated total assets, net assets and equity (ROTA, RONA and ROE respectively). In view of our interest in the role of debt and equity in financial structure several measures of gearing are calculated: first a simple stock measure expressing all loans as a percentage of total assets; second a ratio designed to capture the relative importance of short- and long-term gearing (long-term loans as a percentage of all loans); third a stock measure which relates all loans to shareholders interests; and finally a flow measure showing interest expense as a percentage of earnings before tax. A number of conclusions may be drawn from Table 2.7 and are listed in the following subsections.

Asset structure

1 Small companies have a relatively low ratio of fixed to total assets. In the non-manufacturing industries, for instance, small companies hold 30.9 per cent of their total assets in the form of fixed assets. For large companies the figure is 59.2 per cent.

Table 2.7 The balance sheet structure, gearing and profitability of large and small UK companies in manufacturing and non-manufacturing industries (excluding oil) in the period 1987–9

	Manufacturing companies		Non-manufacturing companies	
	Small	Large	Small	Large
	(Average % 1987–9)			
1. Fixed assets				
Net tangible assets	30.2	34.2	26.4	52.4
Intangibles	0.4	3.3	0.9	3.4
Investments	0.9	7.0	3.5	3.4
Total net fixed assets	31.5	44.4	30.9	59.2
2. Current assets				
Stock and work in progress	19.6	19.8	25.0	14.0
Trade and other debtors	37.9	23.6	36.0	19.3
Investments	1.0	2.8	0.6	1.7
Cash and short-term deposits	9.9	9.3	7.9	5.8
Total current assets	68.5	55.6	69.1	40.8
Total current and fixed assets	100.0	100.0	100.0	100.0
3. Current liabilities				
Bank overdrafts and loans	11.3	6.1	11.0	4.4
Directors short-term loans	0.5	0.0	2.7	0.0
Other short-term loans	0.3	1.1	1.1	1.3
Trade and other creditors	35.3	23.6	41.9	21.9
Dividends and interest due	0.3	1.7	1.3	1.5
Current taxation	7.1	4.8	4.4	3.9
Total current liabilities	55.0	37.3	62.4	33.1
Net current assets	13.5	18.3	6.8	8.3
Total net assets	45.0	62.7	37.6	67.5
4. Long-term liabilities				
Shareholders interests[1]	36.1	42.0	26.8	47.4
Minority interests and provisions	2.3	6.5	1.1	2.8
Loans[2]	3.2	11.5	6.3	15.4
Other creditors and accruals	3.4	2.7	3.5	1.9
Total capital and reserves	45.0	62.7	37.6	67.5
Total capital and liabilities	100.0	100.0	100.0	100.0
5. All loans as % total assets	15.3	18.7	21.1	21.1
All loans as % shareholders interest	42.4	44.3	78.7	44.5
Long-term loans as % all loans	20.5	61.7	29.4	72.9
Interest expense as % earnings before tax	15.4	12.4	21.4	14.7
6. Pre-tax return on net assets	15.9	19.6	19.1	14.4
Pre-tax return on total assets	12.4	14.3	13.0	17.6
Pre-tax return on equity	10.4	19.1	18.8	13.3

Source: Business Monitor MA3 *Company Finance*: Various Issues
Notes: 1 Ordinary, plus preference plus capital and revenue reserves.
2 Directors loans, bank loans, convertible and debenture loans, all of which have a duration of over one year.

2 In keeping with our modified POH we find small companies have a relatively high proportion of trade debt in their asset structure. Thus in manufacturing trade and other debtors comprise 37.9 per cent of total assets. For large manufacturing companies the figure is 23.6 per cent.

3 There is, however, little to choose between large and small firms in the extent to which they hold liquid assets such as cash or investments, which is not in line with the POH prediction.

Current liabilities

1 Current liabilities are a higher proportion of total liabilities for small than for large companies especially in non-manufacturing, where the respective percentages were 33.1 per cent and 62.4 per cent.

2 Small companies are, as predicted, more reliant on short-term bank loans and overdrafts than large companies (4.4 per cent and 11.0 per cent of total assets respectively in the non-manufacturing sector for instance).

3 Trade and other creditors are a higher proportion of liabilities for small than for large companies. In manufacturing trade and other creditors were 35.3 per cent of total liabilities compared to 23.6 per cent for large companies. In comparison with the trade debtor ratios we find that the small manufacturing companies were net receivers of credit. For large companies debtors and creditors cancel out.

Long-term liabilities

1 As predicted we find that small companies are less reliant on shareholders interests to finance their assets. In manufacturing for instance these accounted for 36.1 per cent of total small firm liabilities compared to 42.0 per cent for large companies.

2 In non-manufacturing we also find as predicted that gearing (as measured by the stock ratio of total loans to shareholders interest, or by the flow measure of interest expense as a percentage of earnings before interest and taxes) is higher for small companies than for larger companies.

3 In manufacturing, small companies are slightly more highly geared on the flow measure and slightly *less* highly geared than larger companies on the stock measure.

4 If we compare the ratio of total loans to total assets, however,

there is very little to choose between large and small companies in non-manufacturing. In manufacturing the larger firms are more highly leveraged using this measure.

5 In both manufacturing and non-manufacturing small firms are much more reliant on short term loans. Thus in manufacturing long term loans are only 20.5 per cent of all loans for small firms, but 61.7 per cent of all loans for large firms.

Thus in manufacturing we find little difference in overall gearing but as the POH would predict within the overall level of gearing a greater emphasis in small firms is placed on short term loans. The basic findings for the asset and liability structures of small companies are thus broadly consistent with the modified POH hypothesis. The high reliance on short term finance provided by banks, and the relatively low proportions of assets financed by shareholders interests are clearly long run persistent features of small business finance. The same is true of the relative importance of trade debt, and trade credit and the relative unimportance of fixed assets in their balance sheet structure. Thus our results match the results of previous investigations for the 1960s and 1970s for the UK and for other countries (HMSO 1971; van der Wijst 1989, Osteryoung, Constand and Nast 1992; Tamari 1980; Holmes and Kent 1991; Hutchinson 1986; Johns *et al.* 1990; Bureau of Industry Economics 1987; Committee of Inquiry on the Australian Financial System 1981, Garvin 1971, Groves and Harrison 1974, Schnabel 1992, Storey *et al.* 1987).

It does appear, however, that the extent of reliance on short-term loans and overdrafts has if anything increased since the Bolton and Wilson Reports where the differences reported between larger and small firms are somewhat less than those shown in Table 2.7, a trend which has been remarked upon elsewhere (Burns and Dewhurst 1987). We can probe this issue further by making use of our company panel data set covering the years 1977–83. These data relate only to companies which were alive and independent throughout this period. For a discussion of the financial and other characteristics of non-survivors see Cosh and Hughes (this volume, Chapter 8). In using this data we provide a finer disaggregation by size. Thus Table 2.8 shows total loans as a percentage of total assets for six size classes with *lower* size class limits of £50,000, £100,000, £500,000, £2.5 million, £12.5 million and £62.5 million. The data are shown as averages for the years 1977–9, and therefore correspond to a period around ten years prior to Table 2.7. The results shown in the first two rows of the upper section, the frequency table section and the

Table 2.8 The analysis by total asset size class of the ratio of all loans/total assets (%) for a sample of UK companies in the period 1977–9 (Size Groups Measured by Total Assets in 1976)

	All	Size 1	Size 2	Size 3	Size 4	Size 5	Size 6
Total no.	1214	116	241	111	172	325	249
Complete data	1214	116	241	111	172	325	249
Mean	15.63	14.27	13.69	13.99	13.92	16.29	19.18
Median	13.52	7.73	10.13	9.76	12.15	14.38	18.48
St deviation	14.15	19.17	14.30	16.53	12.10	14.17	10.30
Skewness	2.54	2.53	1.46	2.88	1.92	4.06	0.80
Kurtosis	15.94	9.31	2.29	13.78	6.44	35.94	1.06
Minimum	0.00	0.00	0.00	0.00	0.00	0.00	0.00
Lower quartile	5.57	0.00	2.36	1.72	5.02	7.01	12.00
Upper quartile	22.27	21.34	20.36	20.21	19.66	22.21	25.66
Maximum	164.52	123.47	72.31	119.89	79.70	164.52	54.48

Frequency Table – percentage of companies falling to each range

Ranges based on all	Size 1	Size 2	Size 3	Size 4	Size 5	Size 6
Minimum to lower quartile	44.83	36.10	35.14	27.91	18.46	6.83
Lower quartile to median	16.38	23.65	27.03	27.91	28.31	23.29
Median to upper quartile	16.38	17.84	16.22	24.42	28.31	35.74
Upper quartile to maximum	22.41	22.41	21.62	19.77	24.92	34.14

Significance Tests

		Size 2	Size 3	Size 4	Size 5	Size 6
+ row av. > column av.	Size 1	+	+	+	–	– – –
+ + sig. at the 5% level		–	–	–	– –	– – –
+ + + sig. at the 1% level	Size 2		–	–	– –	– – –
– row av. < column av.			+	–	– –	– – –
– – sig. at the 5% level						
– – – sig. at the 1% level	Size 3			+	–	– – –
				–	– –	– – –
Upper entry in each cell is the test of means and the	Size 4				–	– – –
lower entry is the test of					–	– – –
medians					–	– – –
	Size 5				–	– – –

significance tests section show that it is only the very largest firms in size class 6 (and to a lesser degree those in size class 5) which have a significantly higher ratio of long-term loans to total assets, otherwise the results for this ratio are much the same as Table 2.9. More interestingly Table 2.9 reveals the extent to which these gearing patterns are altered if we focus on long-term loans alone. Thus we

Table 2.9 The analysis by total asset size class of the ratio of all long-term loans/total assets (%) for a sample of UK companies in the period 1977–9 (Size Groups Measured by Total Assets in 1976)

	All	*Size 1*	*Size 2*	*Size 3*	*Size 4*	*Size 5*	*Size 6*
Total no.	1214	116	241	111	172	325	249
Complete data	1214	116	241	111	172	325	249
Mean	4.91	3.99	2.90	3.34	3.61	5.83	7.70
Median	1.51	0.00	0.00	0.00	1.92	2.51	6.14
St deviation	9.54	14.18	8.03	11.04	4.43	10.98	6.97
Skewness	6.27	5.52	4.13	7.49	1.77	7.05	1.63
Kurtosis	64.74	34.49	19.48	64.27	3.89	76.17	3.31
Minimum	0.00	0.00	0.00	0.00	0.00	0.00	0.00
Lower quartile	0.00	0.00	0.00	0.00	0.00	0.38	2.84
Upper quartile	6.69	0.48	1.25	2.05	5.34	7.30	10.55
Maximum	143.25	114.51	58.16	105.40	25.72	143.25	39.64

Frequency Table – percentage of companies falling in each range

Ranges based on all	*Size 1*	*Size 2*	*Size 3*	*Size 4*	*Size 5*	*Size 6*
Minimum to lower quartile	0.00	0.00	0.00	0.00	0.00	0.00
Lower quartile to median	80.17	75.93	72.97	43.02	40.92	17.27
Median to upper quartile	7.76	12.03	10.81	37.79	31.08	34.94
Upper quartile to maximum	12.07	12.03	16.22	19.19	28.00	47.79

Significance Tests

		Size 2	*Size 3*	*Size 4*	*Size 5*	*Size 6*
+ row av. > column av.	Size 1	+	+	+	–	– – –
+ + sig. at the 5% level		n/a	n/a	n/a	n/a	n/a
+ + + sig. at the 1% level						
	Size 2		–	–	– – –	– – –
– row av. < column av.			n/a	n/a	n/a	n/a
– – sig. at the 5% level						
– – – sig. at the 1% level	Size 3			–	– –	– – –
				n/a	n/a	n/a
Upper entry in each cell is						
the test of means and the	Size 4				– – –	– – –
lower entry is the test of					–	– – –
medians						
	Size 5					– –
						– – –

find in row two of the upper section of Table 2.9 that for firms with total assets up to £2.5 million the median holding of long-term loans is zero, at least 50 per cent of these firms have *no* loans with more than one year's maturity. The mean which is a less satisfactory measure of central tendency given the skewness of the distribution

Table 2.10 The analysis by total asset size class of the ratio of all short-term loans/total assets (%) for a sample of UK companies in the period 1977–9 (Size Groups Measured by Total Assets in 1976)

	All	Size 1	Size 2	Size 3	Size 4	Size 5	Size 6
Total no.	1214	116	241	111	172	325	249
Complete data	1214	116	241	11	172	325	249
Mean	10.72	10.28	10.79	10.65	10.31	10.47	11.48
Median	8.16	3.42	7.58	7.70	8.28	8.72	9.36
St deviation	10.60	13.33	11.22	11.83	11.19	9.12	9.34
Skewness	1.64	1.51	1.15	1.47	2.50	1.59	1.62
Kurtosis	3.98	2.11	0.83	1.83	9.95	4.38	3.80
Minimum	0.00	0.00	0.00	0.00	0.00	0.00	0.00
Lower quartile	2.42	0.00	1.07	1.03	2.29	3.72	4.75
Upper quartile	15.58	17.51	16.89	15.80	13.91	14.84	15.72
Maximum	79.70	63.48	55.32	49.58	79.70	63.73	54.48

Frequency Table – percentage of companies falling in each range

Ranges based on all	Size 1	Size 2	Size 3	Size 4	Size 5	Size 6
Minimum to lower quartile	47.41	31.54	32.43	25.58	19.38	11.65
Lower quartile to median	11.21	21.16	20.72	23.26	29.54	32.53
Median to upper quartile	12.07	18.67	21.62	30.23	28.62	30.12
Upper quartile to maximum	29.31	28.63	25.23	20.93	22.46	25.70

Significance Tests

		Size 2	Size 3	Size 4	Size 5	Size 6
+ row av. > column av.	Size 1	–	–	–	–	–
+ + sig. at the 5% level		–	–	–	–	–
+ + + sig. at the 1% level						
	Size 2		+	+	+	–
– row av. < column av.			–	–	–	–
– – sig. at the 5% level						
– – – sig. at the 1% level	Size 3			+	+	–
				–	–	–
Upper entry in each cell is						
the test of means and the	Size 4				–	–
lower entry is the test of					–	–
medians						
	Size 5					–
						–

of this leverage ratio, shows for what it is worth an increase with size when we compare groups 4–6 with groups 1–3, but overall a non-linear pattern emerges with gearing first falling then rising with size.

Finally Table 2.10 shows the ratio of short-term loans, including overdrafts, to total assets. This reveals much smaller differences than in Table 2.7. The frequency table shows that a higher proportion of smaller companies than larger have values for this ratio which lie above the upper quartile. Moreover, the mean values shown in the first row of the upper portion of the table suggest some decrease with size but this is not confirmed by the median, and none of these differences are statistically significant.

Compared to Table 2.7 this suggests that smaller companies in the UK have increased the extent to which they rely on short term loans and overdrafts since the late 1970s (see also Bannock and Doran 1991). Two important caveats should however be borne in mind. First the data in Tables 2.8–2.10 refer to surviving companies only and we know that non-survivors had higher short-term indebtedness than the average (Cosh and Hughes (this volume, Chapter 8)). By contrast the data in Table 2.7 do not relate to a continuing panel. They are an average of separate years each of which contains companies which may subsequently have failed. This may therefore bias the indebtedness ratios upwards compared to Tables 2.8–2.10. Second the data in Table 2.7 are based on 'modified' accounts for smaller companies which require considerable estimation before consistent ratios may be estimated. They may therefore be less reliable than those in Tables 2.8–2.10 which pre-date the widespread introduction of modified accounts.

Profitability

The implied relatively high reliance on internal funds makes assessment of profitability important in determining the growth potential of smaller firms who wish to expand without ceding control. Table 2.7 shows that;

1 The profitability of small manufacturing companies is below that of large manufacturing companies on each of the three measures shown. Thus for example the rate of return on total assets (ROTA) was 12.4 per cent for small, and 14.3 per cent for large, companies.
2 The profitability of small non-manufacturing companies was above that for larger companies when measured as the return on net assets (RONA) or equities (ROE), but below that for larger companies on a ROTA basis. They earned a ROTA of 13.0 per cent compared to the 17.6 per cent ROTA of larger companies.
3 Profitability estimates based on ROTA are relatively low for small firms because of their much higher reliance on short-term liabilities in the form of trade credit. Netting out short-term liabilities to

calculate returns on net assets inevitably raises estimated small company profitability compared to large companies.

These results imply that the relative profitability of large and small firms has been reversed since the Bolton Report.

The Bolton Report (HMSO 1971) provided estimates of ROTA, RONA and ROE for small and large businesses in 1968. These are shown in the first panel of Table 2.11, whilst the second panel shows estimates of RONA for 1973–5 prepared for the Wilson Committee (HMSO 1979). It is apparent from row c that small firm profitability is between 8 per cent and 30 per cent higher than larger company profitability, depending on the measure and time period. This superiority was also recorded in a number of other studies surveyed by the Bolton research team for the period 1958–68 (Tamari 1972, Table 2.21, p. 29). It is apparent, however, that over the years shown the gap was narrowing, and that, as we have seen, it was reversed by the late 1980s.

Once again we can probe a little further behind these profitability figures by making use of our panel data set covering the years 1977–82 which in part fill the gap in time between the profitability results of Tables 2.7 and 2.11. This data set has a number of advantages. First, it allows us to calculate small firm profit rates gross of directors' emoluments which may be arbitrarily separable from 'profits' in the smallest tightly controlled companies. Second, it allows us to distinguish between different time periods 1977–9 and 1980–2 so as to pick up any greater volatility in small business

Table 2.11 The relative profitability performance of large and small firms in the UK in 1968 and 1973–5

| | | 1968 | | | RONA[1] | |
	RONA	ROTA[1]	ROE[1]	1973	1974	1975
a Small[2]	17.8	11.2	18.7	21.0	18.2	16.1
b Large[3]	13.5	9.5	16.5	18.3	16.3	14.9
c as % of b	131.9	117.9	113.3	114.8	111.7	108.1

Source: HMSO (1971) Table 4.IV, p. 45; HMSO (1979) Table 4.6, p. 60
Notes: 1 RONA = Pre Tax Return on Net Assets, ROTA = Pre Tax Return on Total Assets, ROE = Pre Tax Return on Equity.
2 Small firms in 1968 are those businesses employing less than 200 individuals in manufacturing and selected services who responded to the Bolton Committees Financial Questionnaire. Small Firms in 1973–5 are a sample of 300 incorporated businesses with total capital employed of less than £4 million in 1975.
3 Large firms in 1968 and 1973–5 are companies included in the Board of Trade analysis of large non-financial companies. By the mid-1970s 98 per cent of these companies had net assets of more than £1 million and 75 per cent more than £4 m.

profitability in the face of recession in the early 1980s. Finally it allows us to disaggregate by a wider range of sizes.

For the sake of brevity our discussion is confined to three measures of profitability, the ratio of net profit to total assets; (ROTA); the ratio of net profit plus total directors emoluments to total assets (RODTA); and the ratio of gross profits to sales (GPS). Table 2.12 shows measures of central tendency and dispersion for ROTA averaged over the years 1977–9 for 1,191 continuing companies ranging in size from £30,000 to £12.5 billion in terms of total assets in 1976. The same size classes are shown as for our gearing analysis. The first two rows of the upper section, the frequency table section, and the significance test matrix in the bottom third of the table together reveal that smaller firms in general are less profitable than larger ones, but there is a non-linear relationship between size and profits. Thus the smallest three size classes show a lower RONA than the top three but the difference is only statistically significant (on either the mean or median measure) when sizes 1, 2 and 3 are compared with size class 4. Thereafter average ROTA declines insignificantly with size. These results are supported by the frequency table which shows a much greater percentage of size class 1, 2 and 3 firms falling into the bottom quartile of ROTA values than is the case for size classes 4–6.

The years 1977–9 were relatively buoyant compared with 1980–3. The results of an analysis of ROTA for these years is shown in Table 2.13. The impact of the recession is dramatic. Whereas mean ROTA falls from 9.97 per cent to 4.65 per cent for all companies, by far the most dramatic falls occur in the smaller size ranges, with size class 1 falling for example from 8.77 per cent to 1.75 per cent. The result is that size classes 1 and 2, with an upper limit total asset size of £500,000 in 1976, now have a significantly worse profit performance than size classes 4, 5 and 6, where the lower cut off point is £2.5 million. Size class 3 also has a substantially lower ROTA than the large size classes but the wide dispersion of returns prevents this from being statistically significant. The frequency tables for 1980–2 when compared to that for 1977–9 also shows a disproportionate shift of size class 1–3 companies from the higher to the lower quartiles of the ROTA distribution. Tables 2.14 and 2.15 repeat this exercise but adding back total directors' emoluments to profits and calculating RODTA. The results are striking in that we now find size classes 1–3 to be consistently significantly more profitable than size classes 4–6. Moreover the impact of the recession does nothing to change this, with the largest three size classes suffering greater proportionate

Table 2.12 The analysis by total asset size class of the ratio of net profit/ average total assets (%) for a sample of UK companies in the period 1977–9 (Size Groups Measured by Total Assets in 1976)

	All	Size 1	Size 2	Size 3	Size 4	Size 5	Size 6
Total no.	1214	116	241	111	172	325	249
Complete data	1191	106	232	109	171	325	248
Mean	9.97	8.77	9.33	9.30	11.58	10.37	9.77
Median	9.34	7.72	8.38	7.71	11.01	9.76	9.69
St deviation	7.88	10.38	8.90	8.37	8.03	6.95	6.16
Skewness	0.83	2.56	0.25	1.35	0.35	0.48	0.01
Kurtosis	5.96	14.87	1.84	3.96	1.95	3.21	1.68
Minimum	–26.39	–13.70	–26.39	–6.14	–14.80	–21.72	–12.97
Lower quartile	5.38	3.70	3.93	3.54	6.88	5.85	6.68
Upper quartile	14.11	13.35	14.69	13.13	15.71	14.48	12.81
Maximum	75.46	75.46	39.60	50.32	38.76	44.35	30.09

Frequency Table – percentage of companies falling in each range

Ranges based on all	Size 1	Size 2	Size 3	Size 4	Size 5	Size 6
Minimum to lower quartile	38.68	32.33	34.86	18.13	22.15	16.53
Lower quartile to median	18.87	24.57	21.20	21.05	26.15	30.65
Median to upper quartile	21.70	15.95	21.10	26.90	25.23	35.08
Upper quartile to maximum	20.75	27.16	22.94	33.92	26.46	17.74

Significance Tests

		Size 2	Size 3	Size 4	Size 5	Size 6
+ row av. > column av.	Size 1	–	–	– –	–	–
+ + sig. at the 5% level			0	– –	–	–
+ + + sig. at the 1% level	Size 2		+	– – –	–	–
– row av. < column av.			+	– –	–	–
– – sig. at the 5% level	Size 3			– –	–	–
– – – sig. at the 1% level				– –	–	–
Upper entry in each cell is the test of means and the lower entry is the test of medians	Size 4				+	+ +
					+	+
	Size 5					+
						+

falls. Clearly in the recession smaller firms maintained their directors' emoluments more than they were able to protect their trading margins, which as Table 2.16 and 2.17 reveal reflect both the relative levels of profitability shown by ROTA in both periods and the pattern of relatively greater declines for smaller companies as the recession hit.

Table 2.13 The analysis by total asset size class of the ratio of net profit/ average total assets (%) for a sample of UK companies in the period 1980–2 (Size Groups Measured by Total Assets in 1976)

	All	Size 1	Size 2	Size 3	Size 4	Size 5	Size 6
Total no.	1214	116	241	111	172	325	249
Complete data	1185	109	228	108	169	323	248
Mean	4.65	1.75	3.10	4.16	5.48	5.66	5.70
Median	4.61	1.85	2.96	3.68	6.02	5.20	5.92
St deviation	8.70	9.57	8.38	8.06	9.17	8.63	8.16
Skewness	0.06	0.25	–0.01	0.74	–0.12	–0.24	0.47
Kurtosis	2.38	2.97	1.74	1.05	0.30	2.21	6.07
Minimum	–37.24	–27.15	–31.58	–12.80	–23.05	–37.24	–23.69
Lower quartile	–0.31	–3.42	–1.39	–1.16	–0.46	0.81	1.76
Upper quartile	9.20	6.32	7.58	8.13	10.32	10.60	9.82
Maximum	55.03	43.25	35.47	34.36	29.86	30.03	55.03

Frequency Table – percentage of companies falling in each range

Ranges based on all	Size 1	Size 2	Size 3	Size 4	Size 5	Size 6
Minimum to lower quartile	34.86	33.33	28.70	26.04	19.81	17.34
Lower quartile to median	29.36	26.75	29.63	16.57	26.93	22.58
Median to upper quartile	21.10	20.18	23.15	26.04	24.15	32.26
Upper quartile to maximum	14.68	19.74	18.52	31.36	29.10	27.82

Significance Tests

		Size 2	Size 3	Size 4	Size 5	Size 6
+ row av. > column av.	Size 1	–	– –	– – –	– – –	– – –
+ + sig. at the 5% level		–	–	– – –	– – –	– – –
+ + + sig. at the 1% level						
	Size 2		–	– – –	– – –	– – –
– row av. < column av.			–	– –	– –	– – –
– – sig. at the 5% level						
– – – sig. at the 1% level	Size 3			–	–	–
				–	–	– –
Upper entry in each cell is the test of means and the	Size 4				–	–
lower entry is the test of medians					+	+
	Size 5					–
						–

From the point of view of the finance of company growth these findings highlight the relative financial volatility of small firms to the trade cycle (a volatility which is reinforced by the analysis of company failure provided in Cosh and Hughes, in this volume, Chapter 8). They also confirm the relative riskiness of this sector.

Table 2.14 The analysis by total asset size class of the ratio of net profit and total directors' emoluments/average total assets (%) for a sample of UK companies in the period 1977–9 (Size Groups Measured by Total Assets in 1976)

	All	Size 1	Size 2	Size 3	Size 4	Size 5	Size 6
Total no.	1214	116	241	111	172	325	249
Complete data	945	35	117	84	161	309	239
Mean	13.13	32.99	20.72	14.24	12.39	10.67	9.79
Median	11.33	33.09	20.50	12.38	11.96	10.25	9.75
St deviation	9.68	14.90	11.54	9.53	7.71	6.85	6.07
Skewness	1.37	0.57	0.02	1.43	0.46	0.50	−0.01
Kurtosis	4.50	−0.39	1.22	4.46	1.65	3.71	1.85
Minimum	−21.81	10.95	−21.81	−5.88	−11.96	−21.37	−12.81
Lower quartile	7.39	22.45	13.05	9.05	7.87	6.24	6.77
Upper quartile	16.82	41.25	26.74	18.86	16.56	14.55	12.77
Maximum	68.76	68.76	53.33	58.98	39.15	44.78	30.17

Frequency Table – percentage of companies falling in each range

Ranges based on all	Size 1	Size 2	Size 3	Size 4	Size 5	Size 6
Minimum to lower quartile	0.00	9.40	20.24	23.60	33.01	28.45
Lower quartile to median	2.86	9.40	23.81	23.60	24.92	37.24
Median to upper quartile	11.43	19.66	27.38	29.19	26.86	23.41
Upper quartile to maximum	85.71	61.54	28.57	23.60	15.21	10.88

Significance Tests

		Size 2	Size 3	Size 4	Size 5	Size 6
+ row av. > column av.	Size 1	+ + +	+ + +	+ + +	+ + +	+ + +
+ + sig. at the 5% level		+ + +	+ + +	+ + +	+ + +	+ + +
+ + + sig. at the 1% level	Size 2		+ + +	+ + +	+ + +	+ + +
− row av. < column av.			+ + +	+ + +	+ + +	+ + +
− − sig. at the 5% level	Size 3			+	+ + +	+ + +
− − − sig. at the 1% level				+	+ +	+ + +
Upper entry in each cell is the test of means and the	Size 4				+ +	+ + +
lower entry is the test of medians					+	+ +
	Size 5					+
						+

THE PECKING ORDER HYPOTHESIS AND SMALL BUSINESS FINANCE IN THE UK

In terms of the POH, our results taken as a whole suggest that small firms are characterized by a higher value of liquid to total assets. This

Table 2.15 The analysis by total asset size class of the ratio of net profits and total directors' emoluments/average total assets (%) for a sample of UK companies in the period 1980–2 (Size Groups Measured by Total Assets in 1976)

	All	Size 1	Size 2	Size 3	Size 4	Size 5	Size 6
Total no.	1214	116	241	111	172	325	249
Complete data	880	21	88	80	162	301	226
Mean	7.56	21.96	14.62	8.63	6.40	6.08	5.91
Median	6.86	18.92	14.43	8.14	6.71	5.69	5.96
St deviation	9.39	15.10	11.26	8.02	8.81	8.06	7.97
Skewness	0.62	1.14	–0.57	0.85	–0.10	0.04	0.65
Kurtosis	3.29	1.03	2.03	1.11	0.28	0.97	7.07
Minimum	–28.84	0.91	–28.84	–3.93	–21.85	–25.80	–23.64
Lower quartile	1.77	10.69	8.78	2.50	0.49	1.33	1.93
Upper quartile	12.27	25.03	21.79	12.77	11.47	10.70	9.89
Maximum	64.39	64.39	41.68	37.39	30.33	29.46	55.10

Frequency Table – percentage of companies falling in each range

Ranges based on all	Size 1	Size 2	Size 3	Size 4	Size 5	Size 6
Minimum to lower quartile	4.76	7.95	22.50	30.86	29.37	24.34
Lower quartile to median	9.52	13.64	16.25	19.75	28.71	32.74
Median to upper quartile	14.29	15.91	35.00	28.40	22.11	27.43
Upper quartile to maximum	71.43	62.50	26.25	20.99	19.80	15.49

Significance Tests

		Size 2	Size 3	Size 4	Size 5	Size 6
+ row av. > column av.	Size 1	+ +	+ + +	+ + +	+ + +	+ + +
+ + sig. at the 5% level		+	+ + +	+ + +	+ + +	+ + +
+ + + sig. at the 1% level						
	Size 2		+ + +	+ + +	+ + +	+ + +
– row av. < column av.			+ + +	+ + +	+ + +	+ + +
– – sig. at the 5% level						
– – – sig. at the 1% level	Size 3			+ +	+ +	+ + +
				+	+ +	+ +
Upper entry in each cell is						
the test of means and the	Size 4				+	+
lower entry is the test of					+	+
medians						
	Size 5					+
						–

is consistent with the hypothesis but in term of investments, and cash and short-term deposits, however, Table 2.7 also showed that there was little difference between large and small companies. So the main cause of this higher ratio is a much higher degree of trade debt in current asset structures which poses severe problems in recession.

Table 2.16 The analysis by total asset size class of the ratio of gross profit/ sales (%) for a sample of UK companies in the period 1977–9 (Size Groups Measured by Total Assets in 1976)

	All	Size 1	Size 2	Size 3	Size 4	Size 5	Size 6
Total no.	1214	116	241	111	172	325	249
Complete data	1023	27	159	99	170	322	246
Mean	8.78	5.70	7.75	7.11	9.89	9.32	8.96
Median	7.61	4.39	6.10	5.38	9.06	8.30	8.03
St deviation	7.18	3.91	8.38	7.73	6.62	6.29	7.60
Skewness	2.14	0.80	2.56	1.86	0.82	1.10	3.44
Kurtosis	12.60	–0.21	10.18	5.78	1.73	2.39	27.91
Minimum	–12.76	1.01	–10.39	–11.76	–8.82	–8.44	–12.76
Lower quartile	4.36	2.74	2.64	2.85	5.68	5.05	4.75
Upper quartile	11.81	8.36	10.33	9.46	13.62	12.27	11.77
Maximum	78.38	15.95	56.35	41.85	35.54	38.79	78.38

Frequency Table – percentage of companies falling in each range

Ranges based on all	Size 1	Size 2	Size 3	Size 4	Size 5	Size 6
Minimum to lower quartile	48.15	40.25	36.36	17.65	19.88	19.92
Lower quartile to median	22.22	23.27	29.29	21.76	25.16	26.42
Median to upper quartile	22.22	17.61	15.15	28.82	26.71	29.27
Upper quartile to maximum	7.41	18.87	19.19	31.76	28.26	24.39

Significance Tests

		Size 2	Size 3	Size 4	Size 5	Size 6
+ row av. > column av.	Size 1	– –	–	– – –	– – –	– – –
+ + sig. at the 5% level			–	– –	– –	– –
+ + + sig. at the 1% level	Size 2		+	– –	– –	–
– row av. < column av.			+	– – –	– – –	– – –
– – sig. at the 5% level	Size 3			– – –	– – –	– –
– – – sig. at the 1% level				– – –	– – –	– – –
Upper entry in each cell is the test of means and the lower entry is the test of medians	Size 4				+	+
					+	+
	Size 5					+
						+

Reliance on internal funding therefore means for small firms a heavy reliance on retentions to fund investment flows. The relatively low level of profitability net of directors' emoluments which we find for the small firm sector is coupled (as a separate analysis not reported here shows) with an absence of dividend payment by them and hence a policy of maximum retention for growth and replacement investment.

Table 2.17 The analysis by total asset size class of the ratio of gross profit/sales (%) for a sample of UK companies in the period 1980–2 (Size Groups Measured by Total Assets in 1976)

	All	Size 1	Size 2	Size 3	Size 4	Size 5	Size 6
Total no.	1214	116	241	111	172	325	249
Complete data	963	26	122	95	166	314	240
Mean	5.59	1.61	2.66	4.55	6.48	6.39	6.25
Median	4.66	2.10	3.19	2.93	5.34	5.14	5.34
St deviation	9.73	4.88	15.78	7.89	8.17	9.56	7.37
Skewness	-2.34	-0.09	-5.44	2.05	0.26	1.55	1.39
Kurtosis	47.69	1.48	42.60	8.23	1.08	15.42	6.95
Minimum	-131.31	-10.50	-131.31	-15.00	-22.74	-38.05	-14.75
Lower quartile	1.42	-0.33	1.01	0.14	1.55	2.06	2.16
Upper quartile	9.23	3.93	7.00	6.98	11.41	9.99	9.35
Maximum	83.58	14.74	42.46	47.22	34.58	83.58	52.36

Frequency Table – percentage of companies falling in each range

Range based on all	Size 1	Size 2	Size 3	Size 4	Size 5	Size 6
Minimum to lower quartile	38.46	32.79	36.84	23.49	21.66	20.42
Lower quartile to median	42.31	29.51	25.26	23.49	24.20	22.50
Median to upper quartile	15.38	19.67	20.00	22.29	26.11	31.25
Upper quartile to maximum	3.85	18.03	17.89	30.72	28.03	25.83

Significance Tests

		Size 2	Size 3	Size 4	Size 5	Size 6
+ row av. > column av.	Size 1	–	– –	– – –	– – –	– – –
+ + sig. at the 5% level		–	–	– – –	– – –	– – –
+ + + sig. at the 1% level	Size 2		–	– –	– –	– –
– row av. < column av.			+	–	– – –	– – –
– – sig. at the 5% level	Size 3			–	–	–
– – – sig. at the 1% level				–	– –	– – –
Upper entry in each cell is the test of means and the	Size 4				+	+
lower entry is the test of medians					+	0
	Size 5					+
						–

Thus the median company with total assets of less than £2.5 million paid no dividends at all in the period 1977–9.

Our analysis does confirm the second POH conjecture; although we find little difference in terms of the ratio of overall loans to assets between large and small companies we do find that small firms rely more heavily on short term debt.

Our third POH proposition relating to sources of new funds for expansion is less easily analysed in terms of balance sheet data. We can however turn to the results of the Cambridge National Survey to provide some relevant evidence (for a full discussion see SBRC 1992, Chapter 4).

Sixty-five per cent of the Cambridge sample firms had sought external finance in the previous three years. The proportion in the Bolton sample was 26 per cent but in that survey the question related to the previous two years. However, if we adjust this to 39 per cent in order to make it comparable to our finding, we can see a marked increase in the proportion seeking external finance. It is not possible on the basis of our study to say whether this is a trend or cyclical difference. Moreover the Dun and Bradstreet sampling frame on which the Cambridge survey is based is biased towards firms seeking finance since it is a credit rating database. What is striking, however, is that of this 65 per cent which sought finance only a small number (25) failed to obtain some. This may reflect a reluctance to admit having sought finance and failed to get any or relatively easy financial conditions in the period 1987–90. In any event it contrasts with earlier findings for the UK. Thus in the Bolton sample about a third of those seeking finance failed to obtain any at all and so in that sense there may have been some closing of the small firms finance gap since then.

A comparison of the proportion seeking additional finance across the various groups is shown in Figure 2.1. It can be seen that there is little difference between manufacturing and services, whereas the Bolton survey found that manufacturing firms were far more likely (32 per cent in the previous two years) to have sought additional finance than service firms (20 per cent in previous two years). On the other hand, and like the Bolton survey, the figures show that larger companies and faster growth companies are more likely to have recently sought additional finance. In addition, a higher proportion of newer companies had sought external funding.

Sample firms who had obtained external finance in the previous three years were asked to indicate the proportion obtained from a variety of sources. This gives a clear picture of the relative importance of different sources of finance. The responses are presented in Table 2.18. If we first consider the whole sample we can see that banks are by far the most significant providers of finance to the small firm sector. The banks provided some finance to 84 per cent of those who had raised finance and a separate analysis shows that they accounted for more than half of the additional finance raised for 65

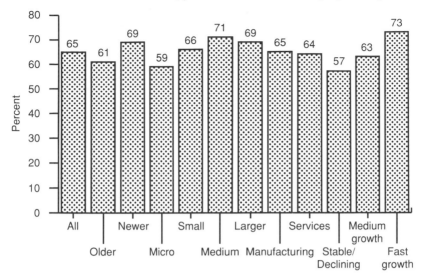

Figure 2.1 Percentage of respondents who sought external finance in the previous three years

per cent of the sample. So in keeping with the POH, loans were dominant as the first source of finance sought after internal cash flow. The mean percentage of additional finance from this source was 61 per cent. The next most important source was hire purchase or leasing which was a source of finance to 45 per cent of the sample firms which had raised new finance and accounted for over half the additional finance in 14 per cent of the cases. This has clearly grown in importance as a significant alternative to further borrowing since the Bolton Report and from the point of view of the contracting parties solves several moral hazard and default problems which would be present if assets were bought on the basis of debt finance. The only other important source was from partners or working shareholders who were identified as a source of finance in 19.5 per cent of the firms and had a mean percentage contribution of 7.6 per cent, again reinforcing the desire to avoid external equity.

When the sample is divided into the different groups the broad picture identified above is maintained, but there is some difference in emphasis across the groups. Newer firms raised a lower proportion of their additional finance from banks and hire purchase or leasing; consequentially, they were more reliant on other sources, particularly private shareholders (both working and other). This reflects the heavy reliance on this sort of finance in the start up phase. In a similar

Table 2.18 Sources of additional finance

Source of finance	All	Micro	Small	Medium	Larger	Older	Newer	Stable/ Declining	Medium growth	Fast growth	Mig.	Serv
% of respondents receiving additional finance from:												
Banks	83.7	80.5	84.1	85.7	90.1	85.4	81.8	82.4	84.7	87.4	85.6	81.6
Venture capital	6.5	2.4	7.1	13.4	6.3	5.5	7.4	4.8	6.0	8.3	7.5	5.4
Hire purchase/leasing	44.6	33.3	50.3	46.2	41.4	50.3	38.8	38.1	51.0	47.8	48.3	40.4
Factoring	6.0	3.7	7.3	7.6	3.6	4.1	7.4	7.3	4.5	5.4	6.2	5.8
Customers/suppliers	8.5	9.8	7.7	7.6	9.9	8.2	8.7	11.7	7.4	8.3	8.9	8.1
Partners/working shareholders	19.5	24.6	17.4	25.2	13.5	15.3	23.7	15.4	18.4	23.7	15.0	24.5
Other private individuals	5.6	9.4	4.9	3.4	1.8	2.5	8.4	4.0	3.3	7.6	4.1	7.2
Other sources	9.7	8.1	9.3	11.8	14.6	9.4	10.0	9.5	9.3	10.1	8.9	10.6
Mean % share by source of finance												
Banks	60.6	60.8	59.1	60.1	68.9	64.6	56.6	63.5	62.1	58.5	62.5	58.4
Venture capital	2.9	1.0	3.4	5.2	2.6	2.4	3.4	2.6	2.4	3.7	3.1	2.8
Hire purchase/leasing	16.0	13.9	18.3	12.6	12.7	17.5	14.6	13.4	18.1	16.8	18.2	13.4
Factoring	3.5	2.2	4.1	4.2	2.4	2.2	4.5	3.9	2.7	3.1	3.3	3.7
Customers/suppliers	2.2	2.6	2.0	1.9	1.8	1.9	2.5	3.2	1.9	2.3	1.9	2.6
Partners/working shareholders	7.6	11.8	6.7	7.2	3.6	5.5	9.9	7.1	7.0	8.1	4.6	11.1
Other private individuals	1.7	3.0	1.5	0.8	0.4	0.6	2.8	1.1	1.1	1.7	1.4	2.0
Other sources	5.4	4.7	4.9	8.2	7.5	5.2	5.7	5.3	4.8	5.8	4.9	6.1
Total responses (no.)	1185	297	648	119	111	561	598	273	418	278	627	554

way service firms were less dependent on these two main sources and far more reliant on partners and working shareholders.

As might be expected, the fast growth firms appear to have drawn their new finance from a wider variety of sources than slower growing firms. Thus apart from factoring and customer/suppliers, both of which were of minor importance, the fast growing group had a higher proportion from each of the other sources. But even the fast growth group was heavily dependent on bank finance and obtained very little outside equity.

The comparison of size groups shows that larger firms had the highest proportion of bank finance and that the micro firms were more reliant on private individuals, partners and working shareholders. The results also suggest that, whilst finance from venture capitalists is generally insignificant, its greatest impact was among medium-sized firms.

These results confirm the unimportance of outside equity as a source of new finance amongst small firms and even amongst those growing fastest; increasing debt, which we have seen in this period was for small firms primarily short term, was the overwhelmingly most significant route followed by raising additional 'insider' equity capital, much as the POH would suggest.

EQUITY GAPS, DEBT GAPS AND SUPPLY OF FINANCE TO SMALL FIRMS IN THE 1980s

We have seen that in many respects the financial structure of small firms is consistent with a strategy of growth and investment finance directed towards maintaining inside control and freedom from constraints. We have also argued, however, that for firms seeking expansion and who are willing to cede control there may still be problems in principle in both the debt and equity market. What implications follow from our analysis for the nature of equity and debt gaps in the UK? To answer this we must go beyond the bare bones of the financial ratios themselves. We begin with a brief resumé of the 'gap' arguments.

In 1979 the Wilson Committee to Review the Functioning of Financial Institutions concluded in its report on the financing of small firms that they were at a disadvantage in terms of the cost of loan finance and the security often required to obtain it. They felt that this reflected the riskiness upon which we have already commented. However they took the view that excessive caution by banks especially in relation to new businesses meant that they might face credit rationing and a debt gap as a result. They therefore recommend the introduction of a loan guarantee scheme. The Wilson Committee also recognized that external equity was difficult to find for small firms and was often available only on unfavourable terms. They noted that venture capital was especially hard to obtain.

> The present equity gap is now seen as having two consequences – a shortage of initial capital for new start-ups, and a deficiency in the provision of development capital to finance the expansion of established enterprises. The latter is particularly serious for fast growing businesses coming up against gearing constraints. But similar problems also affect slower growing firms who may nevertheless want injection of outside equity or occasional marketability for their shares from time to time.
>
> (HMSO 1979: 9)

Despite these remarks they did not believe that the costs of raising capital as such were unfairly high. Instead they focused their recommendations on ways of improving the flow of the small amounts of capital which the large-scale oriented markets failed to supply. Thus they sought to improve the marketability of small firm equity (by promoting an over the counter market (OTC) in their stocks) and of resurrecting 'Aunt Agatha', the family member who could be relied upon as a source of informal capital for entrepreneurial nephews. The demise of Aunt Agatha (brought to a premature end by the burden of taxation) symbolized the absence of small amounts of 'informal' risk capital which might otherwise be available to small-scale entrepreneurs. As a possible solution they suggested reforms to permit the development of Small Firm Investment Companies (SFIC). These would specialize in portfolios of unlimited companies and be given tax breaks to make them relatively attractive vehicles for savings.

Developments in the 1980s have gone some way to meet these objectives. Firstly the Wilson recommendation for a Loan Guarantee Scheme was put into effect in 1981. Secondly, there has been a switch in the balance of taxation away from direct income and capital taxes towards indirect taxation. This may have stirred Aunt Agatha in her grave, or at least perked up her descendants. Thirdly, on the equity front the Business Expansion Scheme was introduced and ran for ten years from 1983. This offered income tax relief at their highest marginal rate for individuals putting new equity investments in unlisted companies so long as they were held for a minimum of five years. Since one of the ways individuals can invest in BES was via specialist investment funds with portfolios of unlisted companies, this was reminiscent of the Wilson SFIC proposals. Fourthly, the 1980s witnessed a large expansion in the venture capital industry, and the development of OTC markets, first via the formation of Unlisted Securities Market in 1980 and then via the Third Market in 1987. Finally the government via the Enterprise Allowance Scheme sought to encourage small business formation and *inter alia* provided financial support for small firms in innovation via the Special Merit Award for Research and Technology (SMART) Scheme and on the regional front provided Regional Enterprise Grants for firms with fewer than twenty-five employees. There has also been a broad programme of national and locally-based initiatives to foster the provision of information and financial advice.

Many of the issues raised by the last major official inquiry into small firm's financing problems seem therefore to have been addressed. Moreover as we have reported here only a handful of firms who claimed to be seeking additional finance in the period 1987–90 reported that they were unable to get any. Another recent major telephone interview survey conducted by the University of Aston concluded that:

> small firms in Great Britain currently face few difficulties in raising finance for their innovation and investment proposals in the private sector . . . programmes of financial assistance are unlikely to have high additionality, unless the programme is rigorously limited to a 'lender of last resort' role . . . the institutional framework for investment finance for small firms is broadly adequate and not in need of further supplementation by public sector initiatives.
>
> (HMSO 1991: 17).

However the Cambridge Survey revealed that manufacturing firms reported a higher level of all forms of constraint on meeting their business objectives than did service firms, and that both groups rated access to finance the most significant of eleven possible factors constraining them (SBRC 1992: 27). The same survey also asked firms to score various factors in terms of their impact on limiting the introduction of new technology, access to finance easily outscored the rest (see further Moore, in this volume, Chapter 4).

Some recent case studies also raise specific doubts about the funding of fast growth and innovative firms. Thus a recent report by the Advisory Council on Science and Technology concluded that some smaller companies circumvented financial barriers to growth by giving up independence through takeover by a larger company. These would of necessity be excluded from Surveys focusing on independent companies. They, and others, also argued the venture capital market and BES funds were not being focused on particularly high risk or innovative activity, especially in manufacturing (ACOST 1990, Hughes 1992). The ACOST study identified a particular difficulty for firms of between fifty and 500 employees in seeking small sums to maintain a sequence of innovations, and recommended *inter alia* initiatives to redirect the BES scheme and introduce SMART-like competitions to boost support for innovation. In one sense this conclusion is not so much at odds with the telephone survey results reported earlier which focused on very small companies, 58 per cent of which employed less than ten individuals and all of which employed less than fifty (HMSO 1991). Even so amongst the sample firms in that study proposals for innovation were more likely to experience financial constraints. The authors reported that 30 per cent of those firms attempting to simultaneously introduce investment, and product and process innovation, expected to experience difficulties (HMSO 1991: 13).

It seems clear that venture capital funding is skewed towards larger investments. Over 90 per cent of BVCA funds in 1990 were managed by businesses with a minimum investment level of over £50,000, whilst over 80 per cent of funds are in businesses with a minimum level of £250,000. A similar picture emerges for 3i the biggest venture capital investor in small firms. In 1990/91 50 per cent of 3i funds were in investments over £1 million, and only 1 per cent were in investments of less than £50,000. So small sums are in relatively short supply (Hughes 1992).

There has also been a decline in the percentage of venture capital funds committed to what might be considered high-tech sectors and companies. Thus the shares of the amount invested and the numbers of companies financed in computer-related, electronic, medical/ genetics and communication sectors all fell between 1984–6 and 1987–90. Investment in these sectors together averaged 38.1 per cent of total investment in the first period but only 23.8 per cent in the second. Venture capital funding has also switched emphasis away from start ups and expansion towards management buy-outs and buy-ins. Between 1983 and 1986 around 23 per cent of funds were invested in start ups and other early stage financing. This fell to an average of around 12 per cent between 1987 and 1990. The share of funds for expansion fell by 10 percentage points between the same two periods, while the share of buy-outs/buy-ins rose from 33.5 per cent to 57.9 per cent (Hughes 1992). Whilst there may be efficiency gains from the reorganization of existing assets between different managements it is apparent that the contribution of venture capital to small business expansion has been less than might be thought from an examination of the raw totals of venture capital funding as a whole.

The Business Expansion Scheme data reveal similar problems. By 1987–8, two-thirds of the BES funds were in investments of over £1 million with only 5 per cent in amounts of less than £50,000 (Inland Revenue 1990). Manufacturing was increasingly neglected in terms of BES funds in the course of the 1980s. Between 1983/4 and 1987/8 the proportion of BES Funds in Manufacturing fell from 33 per cent to 14 per cent. This compares with manufacturing's 27 per cent share of the numbers registered for VAT and which fall within the scope of the BES Scheme. The BES data for 1988/9 and 1989/90 exclude investments in private rented housing (PRH). The extension of the scheme to these projects in those years effectively meant the end of substantial funding for other sectors. Thus of £412 million invested in 2,442 companies in 1988/89, 80 per cent by number, and 87 per cent by value went into PRH schemes. Between 1987/8 and 1988/9 funding for all other sectors dropped from £240m to £50m, and the number of companies supported from 815 to 508. Within that total the number of manufacturing companies funded dropped from 287 to 154 and the amount invested from £24 million to £14 million (Inland Revenue 1990). Whereas BES investments (excluding PRH) were around 0.4 per cent of gross domestic fixed capital formation (excluding private sector dwellings) in the period 1983–8 this fell to 0.08 per cent in 1988/9. The winding up of the scheme announced in the 1992 Budget was hardly a great loss for manufacturing.

Set against the demise of the BES scheme was the introduction in February 1991 of Support for Products Under Research (SPUR) with the aim of providing £30 million over three years in the form of grants for single company innovative projects for those businesses employing less than 500 people. Specially targeted at the 50–500 group highlighted in the ACOST study SPUR offers firms up to £150,000. This is seen as complementary to the SMART Scheme which supports smaller projects. It is too soon to evaluate its impact.

Taken as a whole the developments in venture capital funding and the BES scheme seem not to have been especially well focused on either the smaller end of the funding spectrum nor on the most innovation-intensive manufacturing sectors. This suggests, as the ACOST study concluded, that a finance gap may still exist in these areas. It is however difficult to argue that there were financial constraints on business formation as a whole in the 1980s or that there is a more pervasive market failure for small firms in the availability of funds at least in quantitative terms. In so far as they rely more than they would wish on short-term bank funding or their own internal resources, the 1980s may have posed particular problems for smaller firms. There was a decline in their relative profitability, and the decade began and ended in recessions which imposed severe strains on bank/small firm relationships. Nevertheless in the course of the 1980s itself the small firm sector massively raised its short-term indebtedness as they enjoyed the fruits of the Lawson boom. This honeymoon period did not, however, last.

CONCLUSIONS

The relative financial structure of small UK companies compared to larger ones displays properties that are consistent with the view that they follow a particular pecking order of financial sources in funding their activities. This structure is, with the exception of a substantial increase in their relative short-term gearing in the late 1980s boom, of long standing. Its principal characteristics are that small firms are relatively more reliant on short-term loans and overdrafts and less reliant on equity finance. Trade credit also plays a more important role on the liabilities side of their balance sheet, and trade debtors a more important role on the assets side. The data on small company profitability, however, suggest a significant change in their relative position compared to previous decades. By the 1980s smaller companies had become less rather than more profitable than larger ones. Although here in particular some caution is due because of the limitations of the quality of the data caused by the reporting exemptions introduced for small business in the 1980s. Moreover, adding back directors' emoluments reverses the profitability position.

The financial structure which characterizes small firms in the UK may reflect the wishes of entrepreneurs as much as constraints placed upon them by suppliers of finance. The theoretical literature and the survey and case study evidence we have reviewed does, however, suggest that a variety of market failures may nonetheless produce problems in terms of both credit rationing and excess demands for collateral in certain areas especially for newly established and innovative firms. This is in spite of the introduction of government backed loan guarantee schemes, and a variety of systems of grants and subsidies. The recommendation of new policies to resolve these problems would however take us beyond the remit of this chapter.

NOTE

1 This chapter arises from the research programme into the Determinants of the Birth, Growth and Survival of Small Businesses at the Small Business Research Centre (SBRC), Cambridge University. The research was supported under the ESRC Small Business Programme by contributions from the ESRC, Barclays Bank, Commission of the European Communities (DG XXIII), Department of Employment and the Rural Development Commission. This support is gratefully acknowledged. The authors are grateful to Simon James, Uma Kambhampati, Marc Taylor and Mark Wilson for valuable research assistance.

REFERENCES

Advisory Council on Science and Technology (ACOST) (1990) *The Enterprise Challenge: Overcoming Barriers to Growth in Small Firms*, HMSO, London.

Aghion, P. and Bolton, P. (1992) 'An incomplete contracts approach to financial contracting', *Review of Economic Studies*, 59, 473–94.

Altman, E. (1984) 'A further empirical investigation of the Bankruptcy Costs Question', *Journal of Finance*, September 1067–89.

Ang, J. (1991) 'Small business uniqueness and the theory of financial management', *Journal of Small Business Finance*, 1 (1) 1–13.

Bannock, G. *et al.* (1991) *Venture Capital and the Equity Gap*, National Westminster Bank, October.

Barton, S. and Matthews, C (1989) 'Small firm financing: implications from a strategic management perspective', *Journal of Small Business Management*, 27, January, p. 107.

Bester, H. (1987) 'The role of collateral in credit markets with imperfect information', *European Economic Review*, 31, 4, 887–99.

Binks, M.R., Ennew, C.T. and Reed, G.V. (1988) *The Survey by the Forum of Private Businesses on Banks and Small Firms*, Forum of Private Business, London.

Binks, M.R., Ennew, C.T. and Reed, G.V. (1990) *Small Business and their Banks*, Forum of Private Business, London.

Bragard, L., Donckles, R. and Michel, P. (1985) *New Entrepreneurship*, University of Liege, Belgium.

British Business (1987) 'Lifespan of businesses registered for VAT' 3, April.

Bureau of Industry Economics (1987) *Small Business Review 1987*, AGPS, Canberra, pp. 15–47.

Burns, P. and Dewhurst, J. (1987) 'Great Britain and Northern Ireland' in Burns, P. and Dewhurst, J. (eds) *Small Business in Europe*, Macmillan, London.

Caird, S. (1989) Durham University Business School Occasional Paper 'A review of methods of measuring enterprise attributes', 8914.

Chell, E. (1985) 'The entrepreneurial personality: a few ghosts laid to rest?', *International Small Business Journal*, 3, 3.

Chell, E., Haworth, J.M. and Brearley, S.A. (1991) *The Entrepreneurial Personality Concepts Cases and Categories*, Routledge, London and New York.

Churchill, N.C. and Lewis, V.L. (1983) 'The five stages of small business growth', *Harvard Business Review*, 83 (3), 30–50.

Collins, O.F. and Moore, D.G. (1979) *The Organization Makers*, Appleton-Century-Crofts, New York.

Committee of Inquiry on the Australian Financial System (1981) *Final Report*, AGPS, Canberra (Campbell Committee 1981).

Cooper, A.C. (1986) 'Entrepreneurship and high technology' in Sexton, D.L. and Smilor, R.W. (eds) *The Art and Science of Entrepreneurship*, Ballinger, Cambridge, MA.

Cosh, A.D. and Hughes, A. (1989) 'Ownership, management incentives and company performance: an empirical analysis for the UK, 1968–80', *Economic Discussion Papers*, School of Economics, La Trobe University, Melbourne.

Cosh, A.D. and Hughes, A. (1994) 'Acquisition activity and the small business' in Hughes, A. and Storey, D.J. (eds) *Finance and the Small Firm*, Routledge, London.

Cosh, A.D., Hughes, A., Lee, K. and Singh, A. (1989) 'Institutional investors mergers and the market for corporate control' *International Journal of Industrial Organization*, March, pp. 73–100.

Cosh, A.D., Hughes, A. and Singh, A. (1990) 'Analytical and policy issues' in Cosh, A.D., Hughes, A., Singh, A., Carty, J. and Plender, J. *Takeovers and Short Termism in the UK*, Industrial Policy Paper no. 3, Institute for Public Policy Research, London.

Deeks, J. (1972) 'Educational and occupational histories of owner managers and managers', *Journal of Management Studies*, 9, May, pp. 123–49.

DeMeza, D. and Webb, D. (1987) 'Too much investment: a problem of asymmetric information', *Quarterly Journal of Economics*, 102, 281–92.

Erritt, M.J. (1979) 'The use of company reports for national statistics' *Statistical News*, 44, February, p. 3–5.

Evans, D.S. and Jovanovic, B. (1989) 'An estimated model of entrepreneurial choices under liquidity constraints', *Journal of Political Economy*, 97 (4), 808–27.

Fama, E.F. and Jensen, M.C. (1985) 'Organizational forms and investment decisions', *Journal of Financial Economics*, 14, 1, 101–19.

Friend, I. and Lang, L. (1988) 'An empirical test of the impact of managerial self-interest on corporate capital structure' *Journal of Finance*, June, pp. 271–81.

Garvin, W. (1971) 'The small business capital gap: the special case of minority enterprises', *Journal of Finance*, May, pp. 445–7.

Goffee, R. and Scase, R. (1985) *Women in Charge: The Experience of Female Entrepreneurs*, Allen & Unwin, London.

Groves, R. and Harrison, R. (1974) 'Bank loans and small business financing in Britain', *Accounting and Business Research*, Summer 1974, pp. 227–33.

Hakim, C. (1989) 'Identifying small fast growth firms', *Employment Gazette*, January, 29–41.

Hart, O. and Moore, J. (1990) 'The firm as a collection of assets', *European Economic Review*, 36, 493–507.

Hirsch, R.S. (1986) 'The woman entrepreneur: character skills problems and prescriptions for success' in Sexton, D.L. and Smilor, R.W. (eds) *The Art and Science of Entrepreneurship*, Ballinger, Cambridge, MA.

HMSO (1971) *Report on the Committee of Inquiry on Small Firms* (Bolton Report) Cmnd 4811, HMSO, London.

HMSO (1979) *Interim Report on the Financing of Small Firms* (Wilson Report), Cmnd 7503, HMSO, London.

HMSO (1991) *Constraints on the Growth of Small Firms*, HMSO, London.

Holmes, S. and Kent, P. (1991) 'An empirical analysis of the financial structure of small and larger Australian manufacturing enterprises', *Journal of Small Business Finance*, 1 (2), 141–54.

Holström, B. and Tirole, J. (1990) 'Corporate control and the monitoring role of the stock market', MIT Mimeo, May.

Hornaday, R.W. (1990) 'Dropping the E-words from small business research', *Journal of Small Business Management*, 28, 4 October, pp. 22–33.

Hughes, A. (1992) 'The problems of finance for smaller business', *Small Business Research Centre Working Paper*, no. 15, Department of Applied Economics forthcoming in Dimsdale, M. and Prevezer, M (eds) *Capital Markets and Company Success*, OUP, 1993, Oxford.

Hutchinson, O. (1986) 'The financial profile of small growth firms' in *Australian Small Business and Entrepreneurship Research: Proceedings of the Third National Conference, Institute of Industrial Economics*, Newcastle, September, pp. 181–94.

Inland Revenue (1990) *Inland Revenue Statistics*, HMSO, London.

Jensen, M.C. and Meckling, W. (1976) 'Theory of the firm: managerial behaviour, agency costs and ownership structure', *Journal of Financial Economics*, October, pp. 305–60.

Johns, B., Dunlop, W. and Sheehan, W. (1990) *Small Business in Australia: Problems and Prospects*, Allen & Unwin, 3rd edition.

Kets de Vries M.F.R. (1977) 'The entrepreneurial personality: a person at the cross roads', *Journal of Management Studies*, February, pp. 34–57.

Lafuente, A. and Salas, V. (1989) 'Types of entrepreneurs and firms', *Strategic Management Journal*, 10, pp. 17–30.

Lewis, C. (1979) 'Constructing a sampling frame of industrial and commercial companies', *Statistical News*, February, 44, pp. 6–11.

Mason, C.M. (1989) 'Explaining recent trends in new firms formation in the UK. Some evidence from South Hampshire', *Regional Studies*, 23, 331–46.

Moore, J. (1992) 'The firm as a collection of assets', *European Economic Review*, 36, 493–507.

Myers, S.C. (1984) 'The capital structure puzzle', *Journal of Finance*, 39, 575–92.

Myers, S.C. and Majluf, N.S. (1984) 'Corporate financing and investment decisions when firms have information that investors do not have', *Journal of Financial Economics*, 13, 187–221.

Nicholson, N. and West, M. (1988) *Managerial Job Change*, Cambridge University Press, Cambridge.

62   *Andy Cosh and Alan Hughes*

Norton, E. (1990) 'Similarities and differences in small and large corporation beliefs about capital structure policy', *Small Business Economics*, 2, 3, pp. 229–45.

Norton, E. (1991) 'Capital structure and small growth firms', *The Journal of Small Business Finance*, 1, 1, 161–77.

Osteryoung, J., Constand, L. and Nast, D. (1992) 'Financial ratios in large public and small private firms', *Journal of Small Business Management*, 30 (3), 35.

Pawley, M., Winstone, D. and Bentley, P. (1991) *UK Financial Institutions and Markets*, Macmillan, London.

Penneck, S.J. (1978) 'The top 1500 industrial and commercial companies', *Statistical News* 43, November, pp. 15–17.

Pettit, R. and Singer, R. (1985) 'Small business finance: a research agenda', *Financial Management*, Autumn, pp. 47–60.

Reid, G.C. and Jacobsen, L.R. Jr. (1988) *The Small Entrepreneurial Firm*, Aberdeen University Press, Aberdeen.

Ross, S.A. (1977) 'The determination of financial structure: the incentive signalling approach', *Bell Journal of Economics and Management Science*, Spring, 23–40.

SBRC (1992) *The State of British Enterprise: Growth Innovation and Competitive Advantage in Small and Medium Sized Firms*, Small Business Research Centre, University of Cambridge.

Schnabel, J.A. (1992) 'Small business capital structure choice', *The Journal of Small Business Finance*, 2 (1), 13–21.

Shaw, C. (1993) 'Patterns of success: twentieth century entrepreneurship in the Dictionary of Business Biography', *Centre for Economic Performance*, Discussion Paper no. 144, London School of Economics, January.

Stanworth, J. and Curran, J. (1973) *Management Motivation in the Smaller Business*, Gower Press, Epping.

Stanworth, J. and Gray, C. (eds) (1991) *Bolton 20 Years On: The Small Firm in the 1990s*, Small Business Research Trust, Paul Chapman Publishing, London.

Stiglitz, J.E. (1972) 'Some aspects of the pure theory of corporate finance: bankruptcies and takeovers', *Bell Journal of Economics and Management Science*, Autumn, pp. 458–82.

Stiglitz, J.E. and Weiss, A. (1981) 'Credit rationing in markets with imperfect information' *American Economic Review*, 71, 393–410.

Storey, D.J. (1982) *Entrepreneurship and the New Firm*, Croom Helm, London.

Storey, D.J. (1990) 'The managerial labour market in fast growth firms' in Joubert, P. and Moss, M. (eds) *The Birth and Death of Companies: An Historic Perspective*, Parthenon, Carnforth, Lancs.

Storey, D.J. and Johnson, S. (1987) *Job Generation and Labour Market Change*, Macmillan, London.

Storey, D.J., Keasey, K., Watson, R. and Wynarczyk, P. (1987) *The Performance of Small Firms*, Croom Helm, Beckenham.

Tamari, M. (1972) *A Postal Questionnaire Survey of Small Firms: An Analysis of Financial Data*, Committee of Inquiry on Small Firms Research Report no. 16, HMSO, London.

Tamari, M. (1980) 'The financial structure of the small firm – an international comparison of corporate accounts in the USA, France, UK, Israel and Japan', *American Journal of Small Business*, April–June, 20–34.

Turok, I. and Richardson, P. (1991) 'New firms and local economic development: evidence from West Lothian', *Regional Studies*, 25, 71–83.

Vickery, L. (1989) 'Equity financing in small firms' in Burns, P. and Dewhurst, J. (eds) *Small Business and Entrepreneurship*, Macmillan, London, pp. 204–36.

Walker, D. (1989) 'Financing the small firm', *Small Business Economics*, 1, 4, 285–96.

Watkins, D.S. (1983) 'Development, training and education for the small firm: a European perspective', *European Small Business Journal*, 1, 3, 29–44.

Wijst, D. van der (1989) *Financial Structure in Small Business: Theory, Tests and Applications*, Springer-Verlag, Berlin, Germany.

Williamson, O.E. (1985) *The Economic Institutions of Capitalism*, Free Press, New York.

3 Informal venture capital in the UK

Colin Mason and Richard Harrison

THE FUNDING CRISIS FOR SMALL AND MEDIUM-SIZED ENTERPRISES

There is growing evidence that a finance gap has emerged in the UK during the recession of the early 1990s in the provision of both debt and equity finance for small and medium-sized enterprises (SMEs). Banks, which are the main source of finance for SMEs in the UK, have introduced some fundamental changes in their lending policy in order to try to rebuild their capital base after significant losses, particularly on their property and small business loan portfolios (Gapper 1993a).[1] This need has been made more pressing by new capital ratio requirements of the Bank of International Settlements (Plender 1993). The effects of these changes in policy can be seen in terms of the availability and cost of bank loans and the conditions under which such loans are made. The potential for banking practice to constrain entrepreneurial performance is particularly acute during recessionary periods when the inherent conservatism of the banks becomes more visible (Ennew and Binks 1993). This is despite attempts, often enforced in rescue situations in larger SME customers, by the banks to consider taking equity stakes as an alternative to extending loan arrangements (Batchelor 1993). Apart from this recent development (which is of very minor significance in the overall scale of bank–SME relationships), there are five key areas where problems in bank–SME relationships have been identified.

First, banks are requiring greater security and more equity (Deakins and Hussain 1991). However, declining property and asset values have reduced the amount of collateral that can be offered as security for a loan, thereby reducing the borrowing capacity of small business owners. Second, banks are giving greater recognition to risk in the pricing of loans. Indeed, as recent research by Vyakarnam and

Jacobs (1991) has suggested, business size and the associated perception of security is an important determinant of bank attitudes in its own right, with bank managers regarding the bigger businesses as better than the smaller businesses. Third, banks have raised charges on services such as cheque clearing and there continue to be claims that they have widened margins on loans (Gapper 1993a). Fourth, banks have limited the access of small firms to overdraft facilities because of the difficulties of recovering much of the money lent in the credit boom of the late 1980s as working capital in the form of overdraft but which was frequently used as a substitute for equity by undercapitalized businesses (Plender 1993; Gapper 1993a; 1993b). The final consequence follows on from this. Having recognized the need for greater information on how loans are to be used and to monitor loans, banks are now undertaking much more detailed investigations of their clients. However, the need for more detailed information, and the costs of obtaining it, will inevitably result in an increase in bank charges. Ironically, therefore, one consequence of an improvement in the bank–SME relationship, regarded as important by many commentators (Ennew and Binks 1993), will be a further increase in the cost of bank-provided finance, and a stimulus to a further deterioration in bank–SME relationships on cost, not relationship, grounds. Greater scrutiny will also mean that it will be much harder for SMEs to use overdrafts and loans as a substitute for equity capital. As a consequence, SMEs will have to seek risk capital from other sources (*Financial Times* 1993a). Many bankers and accountants accept that the consequence of this new attitude amongst banks towards lending 'could force small businesses to seek alternative funds from venture capital funds' (*Financial Times* 1993a: 5).

However, the institutional venture capital industry is not a realistic proposition for the vast majority of SMEs. First, venture capital funds are motivated by capital gains and so will only give serious consideration to fast-growing businesses capable of providing an annual return of 30 to 60 per cent. Second, evaluation and monitoring costs which must preceed and accompany venture capital investments are a significant fixed cost element which makes it uneconomic for funds to make investments of less than about £250,000: initial accountancy and legal fees, for example, are unlikely to be less than £50,000 to £70,000, irrespective of the size of the prospective investment (Batchelor 1993). Third, the focus of the venture capital industry is increasingly on the provision of development capital and management buy-outs (MBOs) and buy-ins (MBIs). Statistics compiled for the British Venture Capital Association (BVCA) indicate that in 1991 MBOs/MBIs accounted for 23 per cent of all financings and 55 per cent of finance invested. By comparison, there were just 273 investments in start ups and early stage ventures (down from 521 in 1989), representing 22 per cent of all financings but just 6 per cent of the total amount invested (BVCA 1992a). A further development is that venture capital funds are encountering difficulties in raising new finance from financial institutions, with the supply of new venture capital funds falling by 70 per cent between 1989 and 1992 (Murray 1990; 1993).

It should be noted that these trends in the availability of loan and equity finance for SMEs are by no means unique to the UK. In the USA there is much talk of a 'credit crunch' for smaller businesses that do not have access to public markets (Peek and Rosengren 1992). Losses on real estate following the burst of the real estate bubble has significantly eroded the capital of banks precisely at the time when regulations concerning their capital requirements had been introduced. In order to conform to regulatory requirements concerning capital-to-asset ratios at a time when their capital has been declining, banks have had to reduce lending. As a consequence, they have been not been able to meet the credit needs of SMEs, most of whom are dependent on banks for finance. Goodman and Allen (1992) note that many entrepreneurial firms in the US are finding it harder to acquire new loans, and are encountering more difficulties in the application/review process, reporting requirements and collateral requirements. Their findings therefore clearly indicate that banks are revoking credit lines on small but credit-worthy mid-market companies. The primary strategy being used by SMEs to cope with this credit tightening is short-term cash flow improvement through techniques such as receivables financing, personal refinancing from savings or refinancing home mortgages and company retrenchment through downsizing, layoffs and salary reductions (Allen and Goodman 1993). However, as the economy rebounds these 'band-aid' solutions will significantly constrain the ability of SMEs to respond and grow.

Meanwhile, the US venture capital industry has also encountered difficulties in raising new funds and is therefore hoarding what finance they have (*Economist* 1991). There has also been a shift in the focus of the industry away from its traditional concern with financing early stage investments in favour of expansions and management buy-outs. This trend is attributed by Bygrave and Timmons (1992) to the displacement of 'classic' venture capital by 'merchant capital' funds. Classic venture capital is concerned with early stage investments and involves skills that add value in company forming, building and harvesting; merchant capital funds, on the other hand, are almost entirely dependent upon institutional investors, emphasize financial engineering know-how, transaction crafting and closing and fee generation, and are obsessed with short-term gains. These trends are being viewed with disquiet in both the UK and the USA. In both the UK and the USA there is growing concern that these changes in the availability of loan and equity finance will adversely affect both the rate of new business formation and the growth of established firms, thereby impairing the ability of the economy to recover from recession (Binks 1993; Allen and Goodman 1993). Many SMEs will require additional finance to cope with increased sales, inventory replacement, plant expansion, equipment purchase and R&D investment as the economy recovers but their ability to raise external finance has been reduced and SMEs' awareness of the various options open to them outside the traditional banking industry (e.g. asset-based financing, private placement, informal investment and strategic alliances) is very limited (Allen and Goodman 1993).

In the UK, the Midland Bank (1992) has highlighted the position of SMEs with already high levels of gearing built up in the late 1980s and weak profitability caused by the recession which will have particular problems in raising finance as the economy picks up. They also note that the difficulty of businesses with insufficient collateral to support new borrowings has been exacerbated by the sharp drop in asset prices in recent years. Without an increase in working capital many companies will be at risk from overtrading to meet increased sales during economic recovery. This danger has been confirmed by a recent survey of companies in North West England by accountants Grant Thornton which noted that over half of the firms in their study needed extra finance for capital spending or working capital (*Financial Times* 1993b). A further outcome is that the overhang of debt, combined with the changes in bank lending practices, have made some businesses reluctant to borrow: there is considerable anecdotal evidence that many new and established SMEs are avoiding bank loans (e.g. see *Financial Times* 1993a).

SOLVING THE SME FUNDING CRISIS: THE ROLE OF INFORMAL VENTURE CAPITAL

These concerns have led to a growing debate about how to tackle this widening finance gap. There is a recognition that the reliance of SMEs on debt must be reduced and that sources of equity finance need to be increased, not least to make the UK's SME sector less vulnerable to the exaggerated boom–bust cycles that characterize the UK economy and which have weakened the sector in comparison with that of other European countries (Plender 1993: 8). There have been a number of proposals for ways in which the supply of equity finance can be increased (for a review see Mason and Harrison 1992). However, many of these proposals, such as the Confederation of British Industry's proposal for the establishment of Local Investment Companies (CBI 1983) have, for a variety of reasons, not been implemented. Meanwhile, those initiatives which have been launched are on a small scale and so are largely ineffectual. Even the Midland Bank's Enterprise Funds initiative, which is one of the more ambitious proposals, seems likely to have only a limited impact on the equity gap in the UK. This initiative involves the establishment of a network of regionally-based development capital funds. The running costs will be subsidized, and professional firms (fund managers, stockbrokers, solicitors, accountants and bankers) will waive up to 60 per cent of their fees to undertake investigative work on serious projects. However, the scheme has found it difficult to raise finance and only three funds are currently operational, against an initial target of thirteen funds. Moreover, the number of investments that are anticipated is limited and because the primary aim of the funds is to provide substantial capital (in the order of £250,000) to help existing businesses to expand (McMeekin 1991) they will not alleviate the shortage of equity capital for start ups and early stage ventures.

Against this background, the 'finance gap' debate in the UK has recently begun to consider the role of the *informal* venture capital market in the financing of SMEs. The informal venture capital market comprises private individuals – commonly referred to as informal investors or 'business angels' – who provide risk capital directly to new and growing businesses in which they have no prior connection. The role and potential of informal venture capital in the UK was first emphasized in a study by the Advisory Committee on Science and Technology (ACOST) in 1990. This report noted that in the USA a well-developed informal venture capital market plays a major role in meeting the financing needs of smaller companies. Indeed, according to Wetzel (1986a: 121), angels 'represent the largest pool of risk capital in the US' and 'finance as many as twenty times the number of firms financed by institutional venture capitalists', while 'the aggregate amount they invest is perhaps twice as big' (Wetzel 1986b: 88). Gaston (1989a) similarly concludes that informal capital is the single largest source of external equity capital for small businesses in the United States, almost exceeding all other sources combined. In terms of the value of investments made, informal investment is almost twice as large as private placements and is eight times larger than professional venture capital investments.[2] Informal investors are particularly important in providing seed and start up financing; according to Freear and Wetzel (1988: 353) 'private individuals are most prominent at the early stages of a firm's development, when relatively small amounts are involved, and in those later stage financings involving under $1 million.' Informal risk capital is also a significant source of external equity capital for small businesses in Canada (Riding and Short 1987a; 1987b; Short and Riding 1989; Government of Ontario 1990). The ACOST report concluded from this evidence that 'an active informal venture capital market is a pre-requisite for a vigorous enterprise economy' (ACOST 1990: 41).

Despite this evidence, the debate on how to tackle the UK's equity gap has been slow to recognize the importance of informal venture capital. Indeed, it was only following the publication of some preliminary research results in late 1991 (Mason *et al.* 1991a), and subsequent media coverage (e.g. Batchelor 1991a; 1991b; Woodcock 1991; 1992a; 1992b; Oates 1992; Outram 1992; Miller 1992; Tirbutt 1993) that the importance of the informal venture capital market has become more widely appreciated, for example, by government ministers (e.g. Forth 1992), business organizations (e.g. Institute of Directors 1992) and the financial community (e.g. BVCA 1992b).

There is no equivalent information to that in the USA on the size of the informal risk capital pool in the UK and its significance as a source of equity capital for SMEs. Recent overviews of the availability of finance for small firms have ignored the role – actual or potential – of business angels in the UK (NEDC 1986; Burns 1987; Boocock 1990). The ACOST study observes that 'we do not know of any study which documents its size' (ACOST 1990: 39) although it suggests that informal investment in the UK remains 'under-developed' by comparison with its US equivalent (ACOST 1990: 41). However, there is evidence that private individuals have emerged during the 1980s 'as an alternative source of finance in Britain for the small company which is unable to raise money from more conventional sources' (Batchelor 1988: 9). This increase in informal investment activity in the UK seems likely to reflect the greater opportunities for wealth accumulation by entrepreneurs and senior managers in industry and commerce for whom an informal investment may be an attractive speculative investment. Until recently, high rates of taxation made it difficult for such people to accumulate sufficient amounts of disposable capital; most was either tied up in their own businesses or saved through tax-efficient institutional channels (ACOST 1990). However, the tendency for salaries of senior employees to increase disproportionately, the increasing use of stock options offering the prospect of capital gain, cuts in the top rate of income tax, 'golden handshakes', generous early retirement incentives to senior managers made redundant, high levels of acquisition of small owner-managed companies by the corporate sector and the creation of the USM and Third Markets to enable entrepreneurs to sell stakes in their companies have all contributed to an increase in the number of business people with disposable wealth. Anecdotal evidence suggests that the small business sector is often viewed as an attractive 'alternative' investment for some of this newly acquired capital (Batchelor 1989; Cary 1993). If US trends are a guide (Conlin 1989), successful entrepreneurs who have sold their companies will be a prime source of business angels.

In the remainder of this chapter we present some findings from what is – to the best of our knowledge – the first study of the informal venture capital market in the UK. In view of this we make no claims that this is a definitive account. Rather, our aim is simply, in Wetzel's phrase, to put 'some boundaries on our ignorance' (Wetzel 1986a: 132). We address five sets of questions:

1 *Investor characteristics*: What are the personal characteristics of informal investors in the UK (e.g. age, education, employment, income and net worth, previous entrepreneurial experience)? How similar, or different are they from US informal investors?
2 *Investment activity*: What is the scale of informal investment activity? How many investment opportunities do they consider? How frequently do they invest? How much do they invest? What information sources are used to identify and evaluate investment opportunities?
3 *Investment portfolios*: What are the characteristics of informal risk capital investments in terms of amounts invested, type of businesses financed (stage, sector, location) and participation with other investors?
4 *Investor involvement in the companies in which they invest*: What is the relationship of investors with the firms in which they invest? Are they active and passive investors? What roles do active investors play? How much voting control do informal investors have?

5 *Investment decision-making processes*: What motivates informal investors? What criteria do they take into account when deciding whether to invest? What are their risk perceptions, reward expectations, and liquidation expectations?

METHODOLOGICAL CONSIDERATIONS

US research has noted that informal investors are extremely difficult to identify. Informal investors have a preference for anonymity, there are no directories of individual investors and no public records of their investment transactions (Wetzel 1981; 1987). They may also be reluctant to respond to research surveys because of the private and personal nature of the subject matter and the fear of being identified and then deluged with investment proposals (Haar *et al.* 1988). The size and characteristics of the population of informal investors is therefore unknown, and probably unknowable (Wetzel 1983). Consequently it is not possible to undertake any survey which is based on a representative sample of informal investors.

Three main approaches have been used by US and Canadian researchers to identify samples of informal investors. However, each of these approaches is problematic. The most common approach has been to undertake large-scale postal surveys, often using purchased mailing lists, of various groups of individuals with a sufficiently high discretionary income that they might be involved in informal investment activity (Wetzel 1981; Haar *et al.* 1988; Myers and Moline 1988; Postma and Sullivan 1990). Examples include: high income investors; presidents of major companies; business school alumni; high income professions (e.g. doctors, dentists, CPAs, accountants, attorneys, brokers and bank CEOs); registered owners of Mercedes Benz cars; and subscribers to business magazines (e.g. INC, In Business, Venture). In addition, some studies have distributed questionnaires through small business associations, professional organizations (e.g. bankers' association, National Association of Securities Dealers), members of venture capital clubs, and financial matchmaking services. However, this approach has a number of difficulties. First, the types of mailing lists used are inevitably a source of bias (e.g. to particular occupations). Second, respondents are likely to be biased towards active and successful investors (Haar *et al.* 1988). Third, it is relatively expensive on account of the costs of postage and purchase of mailing lists. Finally, usable response rates are generally extremely low: studies by Wetzel (1981) in New England (1.3 per cent), Myers and Moline (1988) in Missouri (3.6 per cent) and Haar *et al.* (1988) on the East Coast of the USA (4.3 per cent) all obtained usable response rates of less than 5 per cent. Indeed, having used this approach in his pioneering New England study, Wetzel (1981: 219) observed that 'purchased mailing lists . . . were of limited value in reaching informal investors'.

The second approach, favoured by studies that have been sponsored by the US Small Business Administration, has been to contact informal investors through the firms in which they have invested (Aram Associates 1987; Aram 1989; Gaston and Bell 1986; 1988). This involves sending brochures outlining the nature of the research to the CEOs of a structured random sample of firms on the SBA's database (which is drawn from the Dun and Bradstreet files) with the invitation to pass them on to any informal investors their firm might have. Those investors who indicate that they are willing to participate in the study are then sent a questionnaire. The major advantage of this approach is that it allows national estimates of the scale of the informal risk capital market to be made (Gaston 1989a). However, the disadvantage is that it generates relatively small sample sizes. For example, Gaston and Bell (1986; 1988) obtained 435 usable responses from informal investors from three surveys involving a total of 240,000 small firms. The Aram Associates (1987) study obtained fifty-five usable questionnaires from a survey of 20,000 firms. Two factors account for these low sample sizes. First, a high proportion of the firms that are mailed the brochures will not have raised finance from informal investors, or from any other source of equity finance. Second, the response rate from investors who return the brochure indicating their willingness to participate in the survey is low. Gaston (1989a) reports that only 551 informal investors out of a total of 2,900 who responded via the brochure returned completed questionnaires. Similarly, Aram (1989) noted that in his study only sixty-eight completed questionnaires were returned, of which fifty-five were valid, out of the 200 questionaires that were mailed to informal investors who indicated that they were willing to participate in the study. Moreover, to achieve even this modest response required numerous telephone calls (up to four per investor). The high wastage rate, low response rate and labour-intensive nature of this approach therefore makes it equally problematic.

The third approach is normally described as the snowball, or nominated sample, approach. This approach exploits the fact – first reported by Wetzel (1981) – that investors tend to be linked by friendship and business networks, thus finding one informal investor typically leads to the identification of others. For example, in

Neiswander's (1985) study an initial group of six investors was able to provide referrals to 113 additional investors. These studies have therefore approached local professionals (e.g. lawyers, accountants, bankers), local business organizations (e.g. Board of Trade) and known informal investors in order to identify an initial group of informal investors who have, in turn, identified other informal investors. This approach has mainly been used in small-scale local studies (e.g. Nieswander 1985; Short and Riding 1989), although the postal surveys by Wetzel (1981), Tymes and Krasner (1983), Haar *et al.* (1988) and Postma and Sullivan (1990) have also made use of referrals to complement their mailing lists. Such personalized approaches have achieved high response rates (e.g. 48 per cent by Neiswander 1985; 50 per cent by Short and Riding 1989). Wetzel (1981) notes that personal contact through 'networking' was the most successful method of obtaining completed questionnaires in his New England study. However, because of the research resources required, this approach has generally been based on small sample sizes (e.g. 41 by Tymes and Krasner 1983; 25 by Short and Riding 1989). Problems of representativeness are therefore even more acute than in postal surveys.

The present UK study has involved a combination of the postal survey and 'snowball' sample approaches. A large-scale postal survey was undertaken using six mailing lists, comprising owner-managers of SMEs, high income groups, investors in speculative stocks, contacts of a venture capital fund (Metrogroup) and respondents to an advert in the *Financial Times* by an informal investors syndicate seeking members. In total, over 4,000 questionnaires were sent out.

In addition, questionnaires or brochures describing the project which invited recipients to participate in the study were sent to the following:

- subscribers to *Venture Capital Report* (VCR) – a monthly magazine containing articles about entrepreneurs seeking to raise risk capital (Cary 1993);
- individuals who placed adverts in the *Financial Times* seeking investment opportunities in unquoted companies;
- individuals who were either known to be informal investors or were identified by another individual or organization as informal investors. Some of these individuals were given additional copies of the brochure to pass on to their friends and colleagues.

The questionnaire, a modified version of that used in the US Small Business Administration studies (to facilitate comparison with US

74   Colin Mason and Richard Harrison

Table 3.1 Sources of sample population and response rates

Postal Survey mailing list	A number sent out	B number of usable responses[1]	C number returned by non-investors	D unusable responses[2]	E refusals	F deceased/ gone away
Metrogroup[3,4]	134	9	10	1	1	7
'Your Business' magazine controlled subscription	425	1	93	1	1	1
'OTC/USM investors'	1000	16	240	8	11	12
'Wealthy and professional people'	500	4	128	0	7	4
'Female shareholders'/ 'Wealthy Women'[5]	500	0	88	3	2	3
'2% Capitalists'	500	5	103	1	11	10
'Harvard Securities clients'	1000	17	231	5	2	9
Informal investors syndicate	59	8	8	0	0	1
Sub-Total	4118	60	901	19	35	47

Other Sources:

	number of usable responses
Financial Times adverts	1
Venture Capital Report subscribers[6]	12
Known and referred investors and their contacts	5
Face-to-face interviews	8
Sub-total	26
Grand total	86

Notes:

Overall response rate: $= \dfrac{B+C+D}{A-F} = 24.1\%$

Usable response rate: $= \dfrac{B}{A-F} = 1.5\%$

1 After one reminder letter.
2 Unusable questionnaire responses included those completed by investors in BES funds and prospectus issues, investments made through stockbrokers, investments in public companies, an investment in Traidcraft, and an investment in a theatrical venture.
3 An informal investors syndicate.
4 Pilot survey.
5 Because of the lack of usable responses from the initial mailing no reminder was sent to this group.
6 From 16 subscribers who returned the insert indicating their willingness to participate in the survey.

studies), was piloted on the Metrogroup mailing list and subsequently revised.

The postal survey achieved a response rate of 24.1 per cent (after one reminder letter). However, the vast majority of respondents are not informal investors. The usable response rate, based on 60 completed questionnaires, is just 1.5 per cent (Table 3.1). But clearly, this response rate is in many ways misleading since it does not indicate the proportion of informal investors who have completed the questionnaire; however, as there is no way of knowing the number of informal investors on each mailing list this more appropriate response rate cannot be calculated.[3] By far the most useful mailing lists were those provided by Metrogroup and by the informal investment syndicate (response rates of 7.1 per cent and 13.8 per cent respectively). Of the other mailing lists only the OTC/USM investors and Harvard Securities clients had usable response rates of more than 1.0 per cent. The insert in the VCR magazine has also been quite effective, generating a further twelve usable responses.

This postal survey has been complemented with a number of face-to-face interviews with informal investors in the Hampshire area using the 'snowball' survey methodology. The intention was to obtain the same information as in the postal survey but also to collect supplementary evidence on these and other topics. A number of local organizations were contacted, including local enterprise agencies, banks, accountants, solicitors and economic development organizations. However, only two of these organizations were able to suggest names of informal investors. Interviews with this initial group of five investors (four of whom were provided by one source) led to contacts with a further three investors. Although this approach has achieved a 100 per cent response rate it has proved to be a rather unsatisfactory methodology because of the fortuitous nature of finding initial referrals and the large input of labour required to make contact with individuals, verifying that they are informal investors and undertaking the interviews.

The following discussion is based on responses from eighty-six actual and potential informal investors (seventy-eight self-completed questionnaires and eight face-to-face interviews). It should be clear from the earlier comments concerning the difficulties involved in identifying informal investors that there is no way in which the 'representativeness' of this sample can be assessed. It is also important to note that informal investors are a heterogeneous group. US research indicates that business angels differentiate themselves in a variety of ways. For example, Postma and Sullivan (1990) find that three groups

Table 3.2 Special types of US business angel

Devils	Angels who gain control of the company
Godfathers	Successful, semi-retired, consultants/mentors
Peers	Active business owners helping new entrepreneurs, with vested interest in the market, industry or individual entrepreneur
Cousin Randy	A family-only investor
Dr Kildare	Professionals such as MDs, CPAs, lawyers and others
Corporate Achievers	Business professionals with some success in large corporate organizations but who want to be more entrepreneurial and in top-management roles
Daddy Warbucks	The minority of angels who are as rich as all angels are commonly – and incorrectly – believed to be
High-tech Angels	Investors who invest only in firms manufacturing high-technology products
The Stockholder	An angel who does not participate in the firm's operations
Very Hungry Angels	Angels who want to invest over 100 per cent more than deal flow permits

Source: Gaston 1989b: 10

of informal investors can be identified in terms of their motivation, and Aram (1989) notes that certain types of informal investors have particular commitments to investing in start-up firms and technology-based ventures and to participating with co-investors. Gaston (1989b), perhaps optimistically given the nature of the data available, distinguishes between ten different types of informal investor (Table 3.2). However, our objective here is simply to review the most common characteristics of informal investors in the UK. Although the diversity of informal investors will be highlighted, the limited sample size precludes any detailed examination of the characteristics of sub-groups of the UK's informal investor population.

INVESTOR CHARACTERISTICS

UK business angels are predominately male (99 per cent); there is little evidence in the UK, nor in the USA (Haar *et al.* 1988), that women have entered the ranks of informal investors on any scale. UK informal investors are mainly in middle age (Figure 3.1), with 36 per cent in the 45–54 age range and a further 30 per cent in the 55–64 age range; relatively few are either under 35 years old (3 per cent) or 65 years old and over (13 per cent). In this respect they differ from the typical US informal investor in being significantly older. In Gaston's (1989b) sample of business angels 11 per cent were under

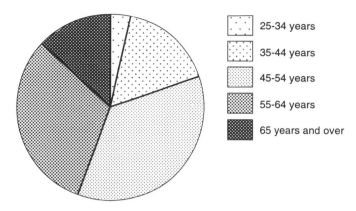

Figure 3.1 Ages of informal investors

Table 3.3 Entrepreneurial background of informal investors

Number of businesses founded	Informal investors	
	No.	%
None	28	32.9
1	17	20.0
2	13	15.3
3	15	17.6
4	4	4.7
5	2	2.4
> 5	6	7.1
No response	(1)	
Total	85	

35 years old, 33 per cent were in the 35–44 year age range and 50 per cent were aged between 45 and 64.

Informal investors in the UK are experienced entrepreneurs, paralleling the US situation (Gaston 1989b). Two-thirds of informal investors have started at least one business and 70 per cent of these have founded more than one business (Table 3.3). Almost two-thirds of these informal investors continue to be connected with at least one of the businesses that they have founded, generally as shareholder and chairman, director or managing director. Nearly half of the business founders have also sold at least one business (42 per cent more than one business), generally either through a trade sale (53 per cent of sales), sale to other investors (26 per cent) or sale to management (20 per cent). Just one informal investor had floated a

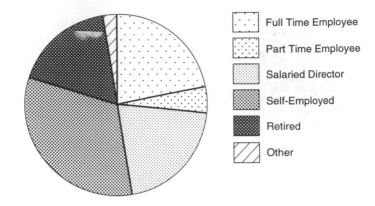

Figure 3.2   Current employment status of informal investors

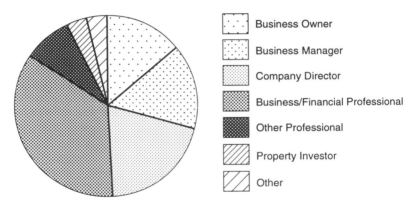

Figure 3.3   Principal occupation of informal investors

business on the stock exchange. Of those respondents who have never founded a business over half (54 per cent) have nevertheless had management experience in a small or medium-sized firm.

The business background of most business angels is further indicated when their current employment status (Figure 3.2) and occupation (Figure 3.3) are examined. In terms of employment status, one-third of informal investors classify themselves as self-employed; this includes those who continue to run their own business as well as others who describe themselves as consultants (which includes a number of investors who have sold their business). A further 22 per cent of informal investors are salaried directors, most of whom are involved in actively managing their own business. Just over one-quarter of informal investors are in paid employment (22 per cent in full-time jobs and 5 per cent in part-time employment) while 17 per cent are retired (Figure 3.2).

An occupational classification indicates that the vast majority of informal investors are business/professionals (accountants, company secretaries, consultants), chief executives/managing directors, company directors and business owners (Figure 3.3).[4] Their management experience is mainly in the areas of finance, marketing, sales and general management (Figure 3.4). In contrast, there are few informal investors amongst science and engineering professionals and medical and educational professionals (e.g. doctors, dentists, teachers, architects). Gaston (1989b) similarly notes that in the USA professional groups outside the business community contain few informal investors.

Informal investors are, not surprisingly, well-off although certainly not 'rich'. In terms of income, 83 per cent had annual incomes of £25,000 and over and 41 per cent had incomes in excess of £50,000. However, relatively few – just 16 per cent – have annual incomes in excess of £100,000 (Figure 3.5a). In similar vein, although the vast majority of informal investors are comfortably off, 54 per cent having a net worth of £250,000 or more (excluding their principal residence), only 19 per cent are millionaires (Figure 3.5b). This situation of most informal investors being well off but with few in the 'super-rich' category also parallels the US situation (Gaston 1989b). We do not have any information on the source of wealth of informal investors but in view of the high proportion of informal investors who have set up, and in many cases sold, one or more businesses it can be assumed that a substantial proportion – indeed, probably the majority – are

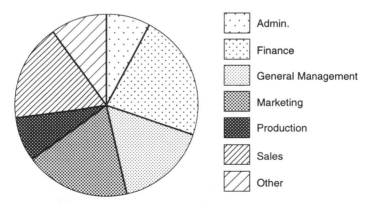

Figure 3.4 Management skills of informal investors

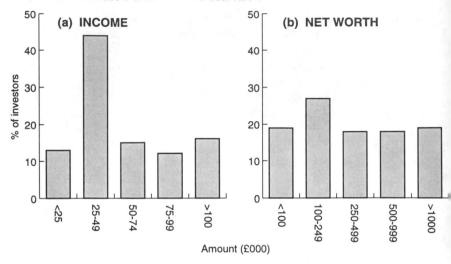

Figure 3.5   Annual household income and net worth of informal investors

financially self-made. In short, informal investment involves 'new' rather than 'old' money.

INFORMAL INVESTMENT ACTIVITY

Our sample of informal investors reported that in aggregate they had come across up to 2,500 investment opportunities during the previous three years. Out of this total they seriously considered around 17 per cent and actually made investments, amounting to £3.1m, in 172 business ventures, equivalent to 8 per cent of all investment opportunities received. This is a higher acceptance rate than that of the formal venture capital industry: in Dixon's (1989) study of thirty venture capital funds only 3.4 per cent of proposals received were funded. Informal investments comprise a mixture of equity funds and loans in approximately a 3:1 ratio. Since loans by informal investors are likely to be more patient than those raised from traditional lenders they are, in effect, quasi-equity.

The 'typical' informal investor has therefore learned of eight investment opportunities in the last three years, of which they seriously considered three (i.e. an average of about one a year) and invested in one business. The median amount invested per investor in the past three years was £22,000. In most cases the investments made by informal investors represent a relatively small proportion of their wealth (Figure 3.6).

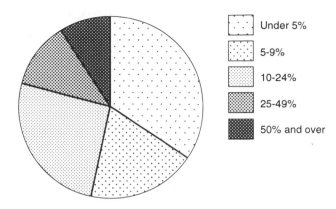

Figure 3.6 Informal investments as a proportion of personal assets

However, the size of the informal risk capital market is substantially larger than indicated by these figures. Seven out of every ten business angels in the sample would have invested more in the last three years if they had come across a greater number of suitable investment opportunities. Indeed, on average, business angels had £50,000 available for investment. In aggregate, the sample of business angels had up to £10 million available for investment, equivalent to three times the aggregate amount that they had invested during the previous three years. It would therefore appear that the informal risk capital market in the UK is a largely untapped source of funds for SMEs seeking venture capital.

These aggregate figures mask considerable diversity between business angels in terms of deal flow and investment activity. First, in terms of deal flow most informal investors received information on less than twenty investment opportunities in the three years prior to the survey. However, a minority (16 per cent) reported receiving information on more than fifty investment opportunities (Table 3.4). In a number of cases this reflects their membership of investor syndicates or financial matchmaking organizations such as LINC and VCR. Second, while the majority of informal investors seriously considered no more than five investment opportunities, 16 per cent of investors seriously considered investing in ten or more businesses (Table 3.5). Most of these investors had received information on a large number of investment opportunities: indeed, they comprised two-thirds of those reporting that they had received more than fifty investment opportunities.

There is rather less diversity in investment frequency. The vast

Table 3.4 Number of investment opportunities received in the three years prior to the survey (1988–90)

Number of investment opportunities received	Number of investors	
	no.	%
< 5	26	32.1
5 – 9	17	21.0
10 – 19	11	13.6
20 – 49	14	17.3
50 – 99	5	6.2
100+	8	9.9
No response	(5)	
Total	81	

Table 3.5 Number of investment opportunities seriously considered in the three years prior to the survey

Number of investment opportunities seriously considered	Number of investors	
	no.	%
None	7	8.8
1	13	16.2
2 – 3	25	31.2
4 – 5	16	20.0
6 – 9	6	7.5
10 – 19	10	12.5
20+	3	3.8
No response	(6)	
Total	80	

majority of business angels are relatively infrequent investors. Nearly one-third of informal investors reported making no investments during the previous three years (although some of these investors had made informal investments before that) and a further 58 per cent made investments in a maximum of three companies (83 per cent of 'active' investors). Just 7 per cent of business angels reported making investments in more than five businesses during the previous three years (Table 3.6). However, there is greater variation between investors in the amounts invested than these figures on investment frequency would imply. While three-quarters of angels have invested under £50,000 during the past three years (79 per cent including non-investors) 20 per cent have invested over £100,000 (14 per cent

Table 3.6 Number of investments made in the three years prior to the survey

Number of investments made in the past three years	Number of investors	
	no.	%
None	26	30.6
1	18	21.0
2 – 3	31	36.5
4 – 5	4	4.7
6 – 9	4	4.7
10+	2	2.4
No response	(1)	
Total	85	

Table 3.7 Total amount invested per investor (equity and loans) in the three years prior to the survey

Amount invested (£000)	Number of investors	%
No investments made	(23)	
< 10	13	22.0
10 – 24	17	28.8
25 – 49	14	23.7
50 – 99	3	5.1
100 – 249	11	18.6
250+	1	1.7
No response	(1)	
Total	59	

including non-investors) (Table 3.7). Although the latter group of informal investors have a higher frequency of investment activity (a median of three investments compared with a median of two investments for those investing less than £100,000) the main distinguishing feature of this group is simply that they have invested a substantially larger amount per firm (£62,500) compared to the other informal investors in our sample (£8,000 per investment).

Variation in the deal acceptance rate amongst informal investors provides a final source of diversity. Regardless of the number of investment opportunities received, the vast majority of angels (82 per cent) have an investment rate of under 50 per cent (Figure 3.7). However, 9 per cent of angels in the sample have invested in all of the business opportunities that have been brought to their attention. In each case these investors have made just one or two investments.

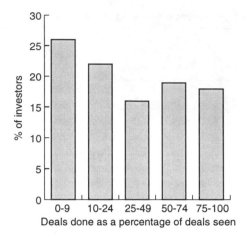

Figure 3.7   Deal acceptance rate

REFERRAL NETWORKS

Business angels derive most of their information about investment opportunities from business associates (cited by 60 per cent of investors) and friends (49 per cent). Although accountants are also a fairly important source of information (cited by 28 per cent of angels), relatively few angels have obtained information on investment opportunities from bankers, stockbrokers and lawyers. One-quarter of informal investors have also identified investment opportunities as a result of their own personal search. However, relatively few business angels have made use of any kind of business referral sources, utilized the services of business brokers or participated in investment clubs or syndicates (Table 3.8).

Referral sources differ in terms of the quality of the investment opportunities that they provide. This can be illustrated in two ways. First, simply examining the referral sources of investments made indicates that the largest proportion have been referred by business associates and friends (Table 3.8). Just over one-third of angels have made investments as a result of information on investment opportunities provided by business associates. Moreover, one-quarter of all investments were brought to the attention of investors through business associates. One-quarter of angels made investments in companies following referrals by friends. This referral source accounted for 16 per cent of all investments made. The third most useful information source was the investor's own personal search for investment opportunities, with 16 per cent of angels reporting

Table 3.8 Referral sources

Referral source[1]	Information received on investment opportunities		Investments made			
	no. of investors	% of investors	no. of investors	% of investors	% of total investments	yield[2]
Informal sources						
Friends	36	48.6	18	24.3	16.3	13
Business associates	45	60.1	26	35.1	24.7	13
Active personal search	20	27.0	12	16.2	15.1	12
Other entrepreneurs	9	12.2	4	5.4	4.2	12
Media	18	24.3	4	5.4	2.4	8
Formal sources						
Accountants	21	28.4	5	6.8	3.6	5
Solicitors	4	5.4	2	2.7	5.4	8
Clearing banks	6	8.1	3	4.1	2.4	15
Merchant banks	7	9.5	2	2.7	4.2	19
Stockbrokers	9	12.2	4	5.4	9.0	31
Organized referral sources						
Investment clubs	3	4.1	1	1.4	1.2	2
LINC	6	8.1	3	4.1	1.8	1
VCR[3]	14	18.9	3	4.1	3.0	1
Business brokers	11	14.9	2	2.7	4.2	4
Other sources	6	8.1	4	5.4	2.4	22

Notes: Based on 74 responses.
1. Most respondents cited more than one referral source.
2. The yield is the number of investments made as a percentage of the number of investment opportunities received from the source. The overall yield is eight (i.e. eight investments occur from every 100 investment opportunities for which information is received.) It should be noted that many responses were approximate: in particular, the yield of the organized referral services is likely to be exaggerated because many respondents indicated that they had received 'hundreds' of investment opportunities.
3. This table is likely to give a misleading indication of the importance of VCR as a referral source because subscribers to VCR were contacted as part of the survey. If responses from this source are excluded, then only 6.8 per cent of investors cited it as a source of information, and only one investor made an investment in a VCR-featured company (1.4 per cent).

investments from this source. This source accounted for 15 per cent of all investments made.

In contrast, very few angels have made investments as a result of referrals from accountants or other professional sources. However, in aggregate these sources account for one-quarter of all investments made by business angels. Even fewer angels have made investments based on information obtained from business referral organizations.

These sources account for just 10 per cent of all investments made by business angels, with business brokers the single most useful organized referral source.

This measure of the quality of each referral source is, of course, closely related to variations in their use. An alternative – and more appropriate – measure of usefulness is indicated by the 'yield' of each referral source – simply the proportion of investments made expressed as a proportion of investment opportunities provided. This provides a very different perspective on the usefulness of various referral sources (Table 3.8, final column). Although friends and business associates, along with personal search and other entrepreneurs, have relatively high efficiency rates (12 or 13 investments resulting from every 100 investment opportunities identified) investors were most likely to invest in investment opportunities that were referred by some of the less frequently used referral sources, notably stockbrokers, merchant banks and clearing banks. Stockbrokers are by far the most efficient referral source, with 31 investments from every 100 referrals, followed by merchant banks and clearing banks with efficiency scores of 19 and 15 respectively. By contrast, solicitors and accountants have relatively low efficiency scores. However, the organized referral sources are the least efficient sources of information on investment opportunities: indeed, LINC and VCR barely generate one investment per 100 investment opportunities while the efficiency of investment clubs and investment brokers is only slightly higher.

INVESTMENT CHARACTERISTICS

The characteristics of the investments made by informal investors can be examined under a number of headings. First, in terms of stage of business development informal investors make investments at all stages of company development with no overriding preference for any particular category (Figure 3.8). However, investments in young companies have been the most frequent, accounting for just over one-third of all investments. Investments in pre-start ups are the least common, accounting for only 9 per cent of all investments made by survey respondents.

Business angels have also made investments in all of the main industrial sectors (Figure 3.9). However, they exhibit a particular interest in the manufacturing sector which accounts for 33 per cent of investments made, nearly half of which were in high-technology firms. When the service sector is also taken into account a total of 37

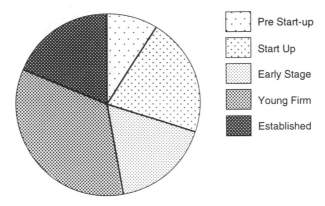

Figure 3.8 Distribution of investments by stage of company development

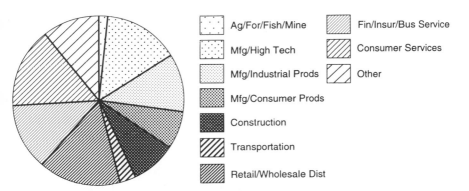

Figure 3.9 Distribution of investments by industry sector

per cent of all investments are in technology-based ventures. Other sectors that have attracted a significant proportion of investments by informal investors are consumer services and retail and wholesale distribution which each accounts for 15 per cent of investments made.

Business angels generally inject very small amounts of capital into the firms in which they invest (Figure 3.10). Just over half of all investments involve amounts of under £10,000 and a further one-quarter of investments involve amounts of between £10,000 and £25,000. At the other extreme, only 12 per cent of investments exceed £50,000. However, about one-third of investments are syndicated between a number of investors (median of four investors), thus the amounts raised by firms are larger than indicated by Figure 3.10.

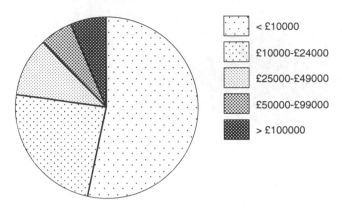

Figure 3.10   Amount invested per company

Nevertheless, the vast majority of informal investments remain well within the 'equity gap'.

These characteristics are in marked contrast to US informal investors (Harrison and Mason 1992a). In comparison with their US counterparts, UK informal investors invest considerably less per firm: the average investment by a US angel (loans plus equity) is just under $60,000 (Gaston 1989b). Furthermore, UK business angels tend to operate more independently. Only 36 per cent have made any investments with other business angels, a much lower proportion than in the US where syndication of investments between a number of business angels (typically friends and business associates) is the norm. For example, Gaston (1989b) reports that only one in twelve business angels is a 'lone wolf' while in Postma and Sullivan's (1990) study of informal investment in Tennessee 72 per cent of angels invested along with other investors personally known to them.

Table 3.9 presents further evidence that most UK informal investors operate independently of other investors. More than two-thirds of respondents indicate that they usually or always make investment decisions independently, and over half stated that their investment decision was rarely or never influenced by recommendations from other investors. Conversely, only a minority of business angels act as lead investors, making independent investment decisions but making recommendations to others. An even smaller proportion are group investors, making investment decisions on the basis of group consensus. US informal investors, by contrast, are much more likely to rely on recommendations from those knowledgeable about the investment opportunity and close personal family and friends (Postma and Sullivan 1990).

Table 3.9 Approach to informal investment activity

	(% of angels)				
	Never	*Rarely*	*Sometimes*	*Usually*	*Always*
Lead investor	35.8	13.4	28.4	13.4	9.0
Independent investor	5.9	5.9	20.6	47.1	20.6
Referred investor	18.5	36.9	26.2	13.8	4.6
Group investor	52.4	17.5	17.5	11.1	1.6

Notes: Lead investor: Someone who actively searches for investment opportunities, makes an independent decision to invest and often suggests investments to others; Independent investor: Someone who welcomes investment leads from others but relies on his/her own investigation in deciding to invest. Other investors do not influence them; Referred investor: Someone who asks questions and reads material but is primarily influenced by a recommendation from a knowledgeable person; Group investor: Someone who invests along with a group of associates. He/she is likely to invest if there is a group consensus, but they do not rely on a single individual. (Definitions based on Postma and Sullivan 1990.)

The majority of business angels are minority shareholders in the companies in which they invest. Just 10 per cent of deals involve a business angel taking a majority stake, while in a further 9 per cent of deals the angel took 50 per cent of the equity. However, in over half of the syndicated deals the informal investor group have acquired majority control.[5] Nevertheless, even with this caveat it is clear that only in a minority of cases does the entrepreneur lose absolute control of the firm to the angel.

Most business angels are also 'hands on' investors. The information in Figure 3.11 has been analysed on the basis of investments because business angels often play different roles in each of the firms in which they invest. Angels have joined the board of directors in 27 per cent of investments, provide informal consulting help on an 'as needed' basis in 21 per cent of cases and the investor works on either a part-time or, less commonly, a full-time basis in the firm in 16 per cent of cases. Thus, investors play a passive role in fewer than one-third of the investments made, merely receiving shareholders' statements and attending shareholders' meetings[6]. A separate study of firms that had raised finance from informal investors (Harrison and Mason 1992b) identified a wide range of support, monitoring and strategic roles played by business angels in their investee companies. Entrepreneurs regarded the angels' role as a sounding board for the management team as being their most valuable 'hands on' contribution (Table 3.10).

Investments by business angels exhibit a strong pattern of geographic localization (Figure 3.12). Just over half of all investments

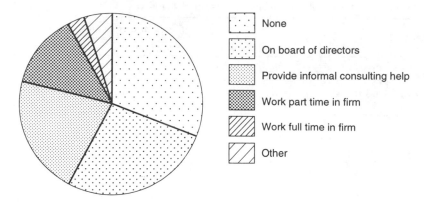

Figure 3.11   Investor participation in investee firms

are made in companies located within 50 miles of the investor's home or office, and two-thirds are in companies located within 100 miles. This geographical pattern is likely to reflect three factors. First, as noted earlier, most informal investors become aware of investment opportunities through informal channels which tend to be highly localized. Second, whereas investors are likely to impose less strict locational limits in considering 'good' investment opportunities, because of the time taken up in the appraisal process only those 'marginal' investment opportunities that are located close to the investor's home/work are likely to be investigated further.[7] Third, and in similar vein, the active role which most investors play in the companies in which they invest is likely to require relatively frequent contact and will discourage investments in companies located some distance away on account of the travelling time involved. US and Canadian studies have also noted that informal investors have a strong preference for making investments in companies located close to their home or place of work (e.g. Wetzel 1981; Gaston 1989b; Short and Riding 1989).

INVESTMENT MOTIVATIONS, CRITERIA AND EXPECTATIONS

Investment motivations

Informal investors are primarily motivated by financial considerations. By far the most important reason given for making informal investments is the opportunity for high capital appreciation, cited by

Table 3.10 Major contributions of informal investors to their investee businesses: the entrepreneur's perspective

Rank role	% of investors playing role	Very helpful	Moderately helpful (percentages)	Not helpful
		Entrepreneur's assessment of investor's contribution		
1 Development of new business strategy to meet changing circumstances	75.0	25.9	37.0	37.0
2 Serving as sounding board to the management team	72.2	53.9	30.8	15.4
3 Monitoring financial performance	66.6	33.3	45.8	20.8
4= Monitoring operating performance	63.9	21.7	47.8	30.4
4= Development of marketing plan	63.9	17.4	43.5	39.1
6= Interface with other members of the investor group	61.1	13.6	54.6	31.8
6= Evaluation of marketing plan	61.1	9.1	50.0	40.9
6= Evaluation of product/ market opportunities	61.1	4.5	59.1	36.4
9= Assistance on short-term crises/problems	58.3	38.1	28.6	33.3
9= Development of actual products/services	58.3	28.6	28.6	42.9
11= Help in obtaining other sources of equity finance	55.6	35.0	20.0	45.0
11= Monitoring personnel	55.6	15.0	35.0	50.0
11= Providing contacts with customers	55.6	10.0	60.0	30.0
14= Development of original business strategy	52.8	22.1	36.8	42.1
14= Replacement of members of management team	52.8	21.1	26.3	52.6
14= Development of product/ service techniques	52.8	10.5	36.8	52.6

Source: Based on Harrison and Mason (1992b)
Note: Only roles which over half of the firms stated that their investors played are listed.

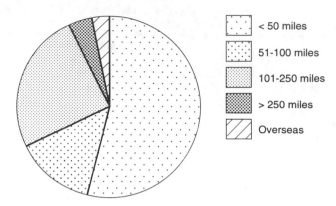

Figure 3.12   Distance of investee firms from investor's home/office

64 per cent of angels as 'very important' and by a further 29 per cent as 'quite important' (Table 3.11). However, there was considerably less unanimity on the importance of other financial considerations as motives for investing. Thus, current or future income was considered to be a 'very important' reason for investing by 29 per cent of informal investors, but 32 per cent considered that this was not an important reason for investing. Similarly, 22 per cent of informal investors gave tax benefits as a very important reason for investing while just under half did not regard this as an important consideration.[8]

US research has reported the importance of non-financial rewards as a motive for investing amongst a significant minority of business

Table 3.11   Reasons for making informal investments

Reason	Very important	Quite important	Not important
For high capital appreciation	64.4	28.8	6.8
Play a role in the entrepreneurial process	39.7	34.2	26.0
For current or future income	29.2	38.9	31.9
For influence over an investment	23.5	50.0	26.5
For tax benefits	22.1	29.4	48.5
For the fun of investing	21.1	39.4	39.4
To help friends/family members	15.9	22.2	61.9
To support new business	14.1	32.8	57.8
To support socially beneficial products or services	10.4	28.4	61.2
For non-financial perks/privileges	3.0	21.2	75.8
For recognition in the community	0.0	4.6	95.4

angels who are prepared to accept lower returns (or assume higher risks) in exchange for psychic income (Wetzel 1981). Many UK business angels are also motivated in part by non-financial considerations. Indeed, the second most important motive for making informal investments was the opportunity to play an active role in the entrepreneurial process which was cited by nearly three-quarters of respondents as being either 'important' or 'very important'. This is likely to reflect the entrepreneurial background of most informal investors which was highlighted earlier. Gaston (1989b) suggests that entrepreneurs hope to re-live their earlier achievements and triumphs through the entrepreneurs that they finance. An alternative explanation is that successful entrepreneurs feel an obligation to support the next generation of entrepreneurs by providing them with finance and experience.

The fun of making informal investments is also an important non-financial motive, being cited by 60 per cent of respondents as either 'important' or 'very important'. However, only a small minority of angels regarded support for socially beneficial products and services, recognition in the community, and non-financial perks and privileges as significant reasons for making informal investments.

Investment appraisal

Informal investors take a variety of factors into account in making their investment decisions. However, just two factors are regarded by a majority of investors as central to the decision whether or not to invest – the management team, cited by 89 per cent of investors, and the growth potential of the market, cited by 71 per cent of investors. Factors which between one-third and half of investors regard as essential considerations are the uniqueness/distinctiveness of the product or service, industry of the business venture, the rate of return, the exit mechanism, the nature of the competition and barriers to competition (Table 3.12).

The importance of 'people' factors, growth prospects and market characteristics in the informal investment decision are underlined by a consideration of the main reasons why angels reject investment proposals. Reasons for rejection are tabulated in two ways: first, angels indicated all reasons for rejecting any investment opportunity; second, angels indicated the two most important reasons for rejecting investment opportunities (Table 3.13). The most significant reason for rejecting investment opportunities was the lack of confidence in the entrepreneur. This was both the most frequent reason given by

Table 3.12 Factors taken into account in making informal investment decisions

	Essential	Moderately essential	Not important
Management team	89.3	8.3	2.4
Growth potential of market	71.4	25.0	3.6
Uniqueness/distinctiveness of product or service	50.0	32.1	17.9
Industry characteristics	45.1	47.6	7.3
Rate of return	43.4	48.2	8.4
Exit mechanism	39.0	42.7	18.3
Nature of competition	37.3	49.4	13.3
Barriers to competition	32.1	44.4	23.5

angels for not investing, cited by 65 per cent of respondents, and also the single most important reason for not investing, cited by 39 per cent of respondents. Another important 'people factor' in the decision not to invest is the lack of expertise of the management team, which was cited by 43 per cent of angels, although by only 4.5 per cent as the most important reason for not investing. Limited growth prospects of the venture was the second main reason for rejecting investment proposals, cited by 56 per cent of angels, while the unattractiveness of the market for the firm's product or service was cited by 49 per cent of angels. However, few angels gave these factors as the single most important reason for not investing. The third major reason for rejecting investment opportunities is associated with equity pricing, with 47 per cent of angels citing the unrealistic value of the equity as a reason for not investing. While this was generally not the most important reason for rejecting an investment opportunity it was nevertheless a very significant secondary consideration, cited by 17 per cent of respondents as the second most important reason for not making an investment. However, failure to agree on the size of equity holding, cited by 27 per cent of angels, was a much less frequent reason for rejecting investment opportunities. A fourth important reason for rejecting investment opportunities was the lack of an obvious exit route.

The vast majority of informal investors normally rely on their own evaluation of investment opportunities either exclusively (40 per cent) or else supplemented by outside advice (54 per cent), typically from their accountant or, less commonly, their solicitor. By contrast, only 5 per cent of investors rely on the expertise of co-investors to evaluate investment opportunities. This serves to further underline the independence of business angels in the UK. However, there is

Table 3.13 Factors in the decision not to invest

	Reason for rejecting investment opportunity	Most important reason for not investing	Second most important reason for not investing
	(% of angels citing this reason)		
No confidence in entrepreneur	65.1	38.8	7.8
Growth prospects limited	55.7	7.5	10.9
Market seemed unattractive	48.6	4.5	7.8
Value of equity unrealistic	47.1	3.0	17.2
Management lacked experience for success	42.9	4.5	12.5
Business concept needed further development	40.0	4.5	1.6
Did not coincide with long-term objectives	35.7	3.0	1.6
Inadequate personal knowledge of firm	34.3	4.5	7.8
Not enough time for appraisal	34.3	4.5	7.8
Lack of information supplied	32.9	3.0	4.7
Not enough knowledge to evaluate firm	30.0	7.5	6.2
Failure to agree on size of equity holding	27.0	3.0	4.7
Unable to assess technical aspects	20.0	–	3.1
Other factors	31.4	11.9[1]	6.2[2]

Notes: 1. No obvious exit route available (10.4 per cent); personality clash with entrepreneur (1.5 per cent); 2. No obvious exit route available.

little consistency in the amount of time which informal investors spend on the serious appraisal of investment opportunities. At one extreme, 39 per cent of investors spend less than 20 hours (18 per cent less than 10 hours) in evaluating an investment opportunity, while at the other extreme 41 per cent devote 20 hours or more (20 per cent more than 30 hours). The remainder of respondents were unable to say how long they spend in evaluating investment opportunities.

Risk perceptions and rate of return expectations

Angels clearly recognize that making informal investments involves a degree of risk which varies depending on the type of firms involved. As Wetzel (1981) notes, the concept of risk poses troublesome problems of definition and measurement. Following his approach, angels were asked to indicate the expected number of 'losers' in a ten-firm portfolio which met their investment criteria. This question elicited a 'don't know' response of between 16 and 21 per cent, which may indicate the limited experience of many respondents in making this kind of investment decision. A further 3 to 6 per cent of respondents stated that the rate of return was 'not important' which further confirms that some investors make informal investments primarily for non-financial reasons. Amongst those respondents who were able to answer this question, pre-start ups are perceived to be the most risky investments. Investments in start ups and early stage ventures are perceived to be slightly less risky while investments in established firms are perceived to be least risky.

However, it is important to note that there is considerable diversity of views amongst informal investors on the level of risk. For example, 47 per cent of angels considered that a ten–firm portfolio of start ups would contain between three and five 'losers'. A further 25 per cent thought that the number of 'losers' would be less than three, while the remaining 28 per cent considered that there would be six or more 'losers'. The perceived risks of investing in pre-start ups and early stage ventures exhibit a similar degree of diversity. There is less variation between investors in perceived risks of investing in young and established firms; indeed, 76 per cent of respondents considered that there would be fewer than three 'losers' in a portfolio of ten established firms. Wetzel (1981) also reports a high level of diversity in risk perceptions amongst informal investors in New England; he suggests that this may simply reflect differences in the level of self-confidence of investors in picking winners but Gaston (1989b) considers this explanation to be unlikely, suggesting instead that it reflects the evaluation of market conditions by investors that they can neither foresee or control.

Variations in the perceptions of informal investors of the risks involved in investing in companies at different stages of development are, in turn, reflected in their minimum rate of return expectations (Figure 3.13). Minimum rates of return decline as the length of period in which the company has traded increases. Thus, informal investors are looking for a minimum rate of return on investment (ROI) which ranges from a median of 45 per cent for pre-start ups and 32 per cent for start ups to 21 per cent for established firms. As a bench-mark, informal investors expect a minimum annual ROI of 15 per cent for investments in 'blue chip' companies. In terms of their expected capital gains, informal investors expect to achieve on average a five-fold increase the value of their original investment over the duration of their holding period.

UK informal investors appear to have slightly higher rates of return expectations than their US counterparts. Gaston (1989b) reports that the median annual ROI amongst US informal investors is 22 per cent.

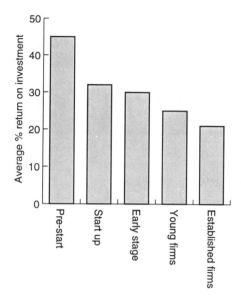

Figure 3.13 Rates of return expectations

They are also higher than those reported by Wetzel (1981) for informal investors in New England, where ROI expectations range from 20 per cent for start ups to 15 per cent for established firms, but are similar to those of informal investors in California (Tymes and Krasner 1983). The capital gains expectations of UK informal investors are also higher than those of US informal investors (Wetzel 1981; Tymes and Krasner 1983).

Exit expectations

US research describes business angels as 'patient investors' (Wetzel 1981). UK informal investors appear to be rather less patient than their US counterparts, although their exit horizon is still greater than that of most venture capital funds. Just over half of informal investors expect to exit from their investments in between three and five years while 23 per cent expect to hold their investments for between six and ten years. A further 9 per cent did not regard the holding period as important and 5 per cent did not have any specific investment horizon (Figure 3.14). This can be compared with an average *maximum* investment horizon of 7.4 years by venture capital funds (Dixon 1989).

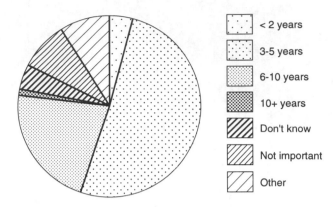

Figure 3.14   Anticipated holding period of investments

There was little unanimity amongst informal investors in how they
were most likely to exit from their investments (Figure 3.15). Indeed,
17 per cent were undecided on their likely exit route, suggesting that
for this group of investors it was not a vital matter in their investment
decision. The most frequently cited exit routes were a sale to, or
merger with, another company and flotation on the stock market,
cited by 29 per cent and 25 per cent of investors respectively.[9] This
emphasis on the sale of equity holdings to 'outsiders' contrasts with
the US informal investors, most of whom expect to sell their shares
to company 'insiders' (Gaston 1989b).

Performance of investments

On balance, informal investors are dissatisfied with the performance
of their informal investments (Figure 3.16). Just under one-quarter of

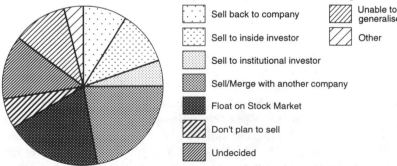

Figure 3.15   Expected method of exit

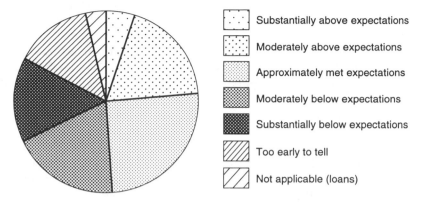

Substantially above expectations

Moderately above expectations

Approximately met expectations

Moderately below expectations

Substantially below expectations

Too early to tell

Not applicable (loans)

Figure 3.16 Satisfaction with performance of investments

informal investors felt that their investments were either performing moderately or substantially above expectations but one-third reported that their investments were performing moderately or substantially below expectations. A further 15 per cent of informal investors reported that it was too early to give an opinion. Compared with US informal investors (Gaston 1989b), a smaller proportion of the investment portfolios of UK investors were performing above expectations and a higher proportion were performing below expectations. Clearly, there are a variety of possible factors which might explain this contrast. One reason is likely to be the higher rate of return expectations of UK informal investors compared to their US counterparts. The onset of recession in 1989, rising numbers of company liquidations and high interest rates are also likely to have influenced both the performance of angels' investments and the returns available from alternative investments, notably savings accounts, which are likely to colour their views of the performance of their informal investments.

CONCLUSION

Two main conclusions arise from this study. First, it clearly indicates that informal investors are playing an important role in the financing of small businesses in UK. Their importance lies in four main areas.

First, business angels represent a major source of risk capital for the small business sector, dwarfing the formal venture capital industry. We estimate that SMEs in the UK have raised about £2 billion from the informal venture capital market.[10] This compares with an investment of about £1 billion per annum by the institutional venture capital industry in the UK, of which well over half is invested in MBOs and established companies seeking development capital (BVCA 1992a). Informal investors typically make investments of under £50,000, and in many cases well below this amount, include both loan as well as equity finance, and their investments are weighted towards new and recently established businesses. Informal investors are therefore making a significant contribution to the filling of the equity gap.

Second, although private investors seek a financial return on their investments in the form of capital appreciation and take account of the quality of management and the growth potential of the business in deciding whether or not to invest, they are, nevertheless, often motivated, in part, by non-financial considerations and so may be more willing to accept lower returns than would an institutional investor. Moreover, they are normally prepared to take a longer-term view of the investment. Further, although they invest in only a small proportion of the investment proposals that they receive, their rejection rate is lower than that of venture capitalists. This reflects the greater willingness of informal investors to invest in an opportunistic way, often without too much study or investigation, particularly if the business is in a market or technology area that they are familiar with.

Third, most business angels play a 'hands on' role in the firms in which they invest. Thus, by raising finance from business angels small firms are able to benefit from the substantial commercial skills, entrepreneurial experience, 'know how' and contacts of their investors. The most valuable contribution made is acting as a sounding board for the management team, and informal investors make more valuable contributions to product and market development than do venture capitalists (Harrison and Mason 1992b).

Fourth, in contrast to the over-concentration of institutional venture capital investments in the South East and adjacent regions (Mason and Harrison 1991), angels seem likely to be much more spatially dispersed throughout the country.[11]

However, it is important to recognize that the informal venture capital market is not a complete solution to the funding difficulties of SMEs. By no means all SMEs either wish to raise equity finance or are attractive to outside investors. Rather, the informal venture capital market is primarily relevant to what in the USA is termed 'high-potential firms' – businesses which grow rapidly and are likely to exceed £10 million in sales – and 'foundation firms' (alternatively termed 'mid-market companies') – businesses which although growing more slowly (10 to 20 per cent a year, compared with over 30 per cent for the high-potential firms) will nevertheless exceed £1 million in sales and may grow to £5 to £10 million. 'Life-style firms', in contrast, rely almost entirely on the personal savings of the owner-manager (Timmons 1990).

In addition, there are some disadvantages associated with raising finance from business angels which should not be ignored. First, although they are generally quite wealthy their pockets are unlikely to be as deep as those of institutional investors and so they may be unwilling or unable to provide subsequent rounds of finance. Second, although most informal investors are happy with a minority stake in the businesses in which they invest they will nevertheless expect a significant say in strategy. Angels who have not made the transition from owner-manager to investor may be particularly meddlesome. And, of course, the clash of two strong egos may create conflict between entrepreneur and angel. Third, some business angels may turn out to be 'devils', with ulterior motives for investing. For example, there is anecdotal evidence of angels who have forced the entrepreneur out of their own firm (e.g. Conlin 1989).

The second main conclusion from this study is that the informal venture capital market is under-utilized by SMEs. Our findings clearly indicate that business angels want to make more investments but cannot find sufficient investment opportunities that meet their investment criteria. As a result, many informal investors have substantial uncommitted funds available – in many cases amounting to two or three times the size of their existing investment portfolio.

This situation reflects three characteristics of the informal investment process. First, most business angels adopt an *ad hoc*, unscientific approach to finding investment opportunities, and place considerable reliance on referrals from friends and business associates. Thus, there is a large degree of serendipity in the number and quality of investment opportunities that come to an informal investor's attention. One consequence of this is that a majority of investors are dissatisfied with their existing referral sources (Mason and Harrison 1992). Second, the search for, and appraisal of, investment opportunities is a time-consuming activity. Few business angels are therefore able to devote significant amounts of time to what is generally a part-time activity. As a consequence, many informal investors have a fairly passive attitude towards making informal investments, awaiting referrals to be 'dropped in their lap' rather than actively seeking out investment opportunities either by exploiting their networks or by undertaking their own search. Third, informal investors exhibit a high level of independence, as reflected in their relatively low level of participation with co-investors. As a consequence, they forgo the opportunity of information sharing. Meanwhile, entrepreneurs are hampered in their search for private sources of venture capital by the anonymity that most informal investors strive to preserve.

Policy implications

The policy implications which arise from this analysis are fourfold. First, in order to increase both the supply of, and demand for, informal venture capital there is a need to raise awareness amongst entrepreneurs, self-made high net worth individuals and professional intermediaries that this form of investment is a financing option. This can be achieved by means of publicity and promotion of informal venture capital in the specialist financial and professional media and through appropriate business and professional associations. Also appropriate in this context is the need for greater guidance to the entrepreneur on how to find business angels and raise finance from them. There are a number of books available to the UK entrepreneur on how to raise venture capital but the content of these publications is exclusively limited to a consideration of formal sources.

Second, there is a need to more effectively mobilize the pool of informal venture capital. Business introduction services seek to over-come the two major sources of inefficiency in the informal venture capital market, namely the invisibility of informal investors and the high search costs for businesses seeking investors and investors seeking investment opportunities, by providing an efficient channel of communication between business angels and entrepreneurs seeking finance. This enables entrepreneurs to bring their investment pro-posals to the attention of a number of private investors simultaneously and provides investors with a convenient means of identifying and examining a range of investment proposals while retaining their anonymity. There is also evidence that business introduction services can contribute to an expansion of the population of active business angels. The number of so-called 'virgin' angels vastly outnumbers active angels. Virgin angels are people who share the same high net worth and business background as active angels but have never made any informal investments. Our evidence suggests that their inability to identify firms which require finance is a major reason for not making informal investments. One of the key factors which might convert virgin angels into active investors is if they received information from a trustworthy source – such as a business introduction service – on companies which wish to raise finance (Mason and Harrison 1993b). There are two main private investor networks in the UK. *Venture Capital Report* (VCR) is a national subscription service with just under 400 private investor subscribers. It publishes a monthly investment bulletin containing fully-researched articles on approximately ten companies seeking finance. Between 1990 and 1992 twenty-four of the companies featured (15 per cent of the total) raised finance from VCR subscribers (Cary 1993).[12] LINC (the Local Investment Networking Company) has been operating since 1987. It is a private sector not-for-profit network financed by corporate sponsorship comprising thirteen independent enterprise agencies in various parts of the country. In addition, a number of other public and private organizations, including enterprise agencies and accountancy firms, also operate match-making services, although in most cases they are small-scale, low-profile and informal.

Match-making services have had only modest success in unlocking the substantial sums of uncommitted money that most business angels have available. This study has noted that very few investors have joined such referral organizations and we have considerable anecdotal evidence from respondents who have been dissatisfied with the quality of referrals that they have received. In the case of LINC, our study has indicated that it has a low awareness amongst informal investors, small business owners and professional intermediaries (e.g. accountants, bankers, solicitors). This reflects the fact that LINC is under-resourced, limiting its marketing and promotion efforts, allied to its patchy geographical coverage. As a result, it has a small client base which has recently stabilized at about 200 investors and about the same number of businesses. More fundamentally, the scale of informal investment activity resulting from LINC's activities has been limited. Only about seventy businesses have raised finance from informal investors to whom they were introduced through LINC since it began in 1987, with the total investment amounting to a little over £4 million, although in 1992 it helped twenty-one businesses to raise £1.14 million, an increase in activity which would appear to reflect the growing difficulties that SMEs are encountering in raising finance from more conventional sources. Participation in LINC has not significantly enhanced the ability of investors to find suitable investment opportunities. Our survey of LINC investors noted that 86 per cent would have invested more if they had come across sufficient investment opportunities. They had an average of £100,000 still available for investment, exceeding the amount that they had invested during the preceding three years by some 40 per cent (Mason *et al.* 1991b; Mason and Harrison 1993c).

The Department of Trade and Industry has recently launched a pilot scheme to part-fund five Training and Enterprise Councils (TECs) to set up private investor demonstration projects. An assessment of their performance is premature in view of the short length of time that they have been operating (but see Mason and Harrison 1993d, for an interim review). However, it seems inevitable that (with possibly one or two exceptions) the impact of these projects in stimulating informal investment activity will also be modest on account of funding limitations and their limited geographical scale of operation.[13]

The third policy implication is the need to make the tax treatment of equity investment in unquoted companies no less advantageous than other forms of saving in order to encourage both active and virgin business angels to make some of their disposable wealth available for investing in smaller companies. Business angels are concerned with capital gains tax: the availability of roll-over relief might therefore be expected to lead to an increase in the willingness of private individuals to invest in unquoted companies (BVCA 1992b). However, it must be emphasized that business angels treat investments on their merits and a tax-break is unlikely to alter their judgement of what constitutes a good and a bad investment opportunity.

Finally, there is evidence that the Financial Services Act (FSA) contains sufficient ambiguity relating to the operation of formal and informal business introduction activities to dissuade professional intermediaries in particular from becoming more significantly involved in *ad hoc* or informal matching. Clarification of FSA requirements in this area could do much to improve the ability of these intermediaries to act as 'archangels', putting together syndicates of informal investors from amongst their clients and contacts to invest in deals that they identify through their commercial activities, a feature which is common in the US informal venture capital market (Spragins 1991). Indeed, the relative absence of this form of informal investment activity may be a signficiant contributory factor to the low level of syndicated deals by informal investors in the UK compared with the USA (Harrison and Mason 1992a).

NOTES

1 Barclays Bank has been writing off bad loans to small businesses at a rate of £1m a day during the second half of 1991 and all of 1992. It made a

pre-tax loss of £242m in 1992–3. In the case of the National Westminster Bank, 47 per cent of its £1.3bn provisions for possible bad debts in its UK branch banking in 1992 were for loans of less than £50,000 (Gapper 1993a).

2 There is now a substantial volume of research on informal investors in the USA: for example, see Wetzel and Seymour 1981; Wetzel 1981; 1983; Tymes and Krasner 1983; Neiswander 1985; Myers and Moline 1988; Haar *et al.* 1988; Gaston and Bell 1986; 1988; Aram Associates 1987; Gaston 1989b; Aram 1989; Postma and Sullivan 1990; Fiet 1991; Sullivan 1991. This literature is reviewed by Mason and Harrison (1993a).

3 Our inability to identify *ex ante* the exact population of informal investors within our sample frames makes it impossible to compare the response rates quoted in Table 3.1 with those reported for other postal surveys where the target population can be clearly identified. In particular, we would expect high non-response rates from those in the sample for whom the topic of the survey is an irrelevance. Furthermore, even for those lists from investment syndicates for which we have crude response rates in excess of 5 per cent, high non-response rates are likely to be characteristic: as we have argued elsewhere (Harrison and Mason 1991), many of those actually involved in these investment syndicates were less than enthusiastic about informal investment, and this is likely to be even more characteristic of those on the syndicate mailing list who chose not to become more actively involved.

If we assume that the response rates for our target, but unidentifiable, population of informal investors (aggregated across all sample frames in Table 3.1) is the same as that for non-investors, we estimate that informal investors constitute about 6.2 per cent of those contacted. If on the other hand we assume that actual informal investors may be up to twice as likely to respond as non-informal investors, this proportion falls to just over 3 per cent (and implies an effective response rate of up to 50 per cent, which is consistent with the experience of the highly targetted survey of Postma and Sullivan 1990). Further investigation of these issues, and the determination of the actual size of the potential informal investment population, is an important issue for further research in this field.

4 Many entrepreneurs classified themselves as managing directors, while a number of ex-entrepreneurs are now self-employed consultants, classified under the business–financial professional category.

5 This may be an overestimate of the proportion of syndicated investments which acquire over 50 per cent of the equity. We suspect that some respondents have given the equity share of *all* of their co-investors (i.e. including company principals) rather than just other informal investors.

6 Although our information on this point is not comprehensive, it is likely that some of these passive investments have been made through the Business Expansion Scheme (BES), a scheme which provided tax incentives for investors making investments in qualifying unquoted companies. Our definition of informal investment includes direct BES investments but excludes investments in BES Funds and in Prospectus Issues (see Mason *et al.* 1988, for further details).

7 This point arose in a meeting of an informal investor syndicate that one of the authors was invited to observe.

8 As before (note 6) we do not have specific information on this point. However, it seems likely that some of the investors who cite tax benefits as a 'very important' or 'quite important' motive for investing have made some or all of their investments through the BES. Indeed, 40 per cent of those investors who cited tax as a very/quite important reason for investing had made BES investments (although these investments were not necessarily their informal investments). Moreover, 61 per cent of BES investors cited tax benefits as very/quite important compared with 47 per cent of non-BES investors.

9 A number of investors gave more than one possible exit route, hence percentages exceed 100.

10 This estimate was based on the following calculation.

(i)   5.45 per cent of companies have raised finance from business angels. (Calculated as an average of studies reviewed in Mason and Harrison 1993a, plus Small Business Research Trust 1991.) This proportion was applied to the stock of companies in Great Britain (Department of Trade and Industry 1992);

(ii) the average investment per investor in a company is £20,000 (the average of the deal sizes reported in Mason *et al.* 1991a and 1991b);

(iii) 39 per cent of deals are syndicated between an average of 3.25 investors (averaged from the figures reported in Mason *et al.* 1991a and 1991b).

11 In view of possible spatial biases in our mailing lists and referrals we cannot reach any definitive conclusion on the regional distribution of informal investors. However, the geographical distribution of respondents to our postal survey does suggest that although informal investors are disproportionately concentrated in the South East, as might be expected, they are nevertheless found throughout the UK. Excluding face-to-face interviews, which have a known southern England spatial bias, and other personal referrals, the regional distribution of respondents is as follows:

South East	41	(56.2%)
East Anglia	4	(5.5%)
South West	3	(4.1%)
East Midlands	2	(2.7%)
West Midlands	9	(12.3%)
Yorkshire–Humberside	3	(4.1%)
North West	1	(1.4%)
North	4	(5.5%)
Wales	1	(1.4%)
Scotland	3	(4.1%)
Northern Ireland	1	(1.4%)
Channel Islands	1	(1.4%)

12 It should be noted that a further 15 per cent of companies featured in VCR raised finance from non-subscribers and a further 17 per cent raised some of the finance that they sought (Cary 1993).

13 The importance of these demonstration projects in increasing awareness of informal venture capital should not be overlooked.

NOTE

This paper has been prepared under the research project 'Informal Risk Capital in the UK' which forms part of the Economic and Social Research Council's (ESRC) Small Business Research Initiative, and is funded by the ESRC in conjunction with Barclays Bank, the Department of Employment, the Rural Development Commission and DG XXIII of the Commission of the European Communities (Ref W108 25 1017). We are grateful to the following individuals and organizations for their various contributions to this study: Jennifer Chaloner, Andrew Blair (Metrogroup plc), Dr Jim Milne (University of Southampton), Dudley Mortelman (Southampton Enterprise Agency), David Nicholas (Impex Southern Ltd), Lucius Cary (Venture Capital Report), Milestone Publications, Media Four Financial, DDM Advertising Ltd (List Broking Division), Dudley Jenkins Group plc and Centaur Direct Response.

REFERENCES

Advisory Council on Science and Technology (ACOST) (1990) *The Enterprise Challenge: Overcoming Barriers to Growth in Small Firms*, HMSO, London.

Allen, K.R. and Goodman, J. (1993) 'The working capital crisis: how rapidly-growing firms are coping', paper presented at the 13th Babson Entrepreneurship Research Conference, University of Houston.

Aram, J. (1989) 'Attitudes and behaviors of informal investors toward early-stage investments, technology-based ventures and co-investors', *Journal of Business Venturing*, 4, 333–47.

Aram Research Associates Inc (1987) *Informal Risk Capital in the Eastern Great Lakes Region*, US Small Business Administration, Office of Advocacy; Washington DC.

Batchelor, C. (1988) 'Private financing: money and time to offer', *Financial Times*, 19 July, p.9.

Batchelor, C. (1989) 'Business angels: an investment of time and money', *Financial Times*, 21 November, p. 21.

Batchelor, C. (1991a) 'Angels give a helping hand to small firms', *Financial Times*, 1 October, p.17.

Batchelor, C. (1991b) 'Flights of pin-striped angels', *Financial Times Venture Capital Survey*, 6 November, p. II.

Batchelor, C. (1993) 'From lender to investor', *Financial Times*, 23 March, p. 15.

Binks, M. (1993) 'Sources of finance for small and medium-sized enterprises in the UK: the banks', paper presented at a CBI Workshop on Finance for SMEs.

Boocock, J.G. (1990) 'An examination of non-bank funding for small and medium-sized enterprises in the UK', *Service Industries Journal*, 10, 124–46.

British Venture Capital Association (BVCA) (1992a) *Report on Investment Activity 1991*, BVCA, London.

British Venture Capital Association (1992b) *Tax Submission 1992/93*, BVCA, London.

108   *Colin Mason and Richard Harrison*

Burns, P. (1987) 'Financing the growing firm', Proceedings of the 10th National Small Firms Policy and Research Conference, Cranfield School of Management, Cranfield Institute of Technology.
Bygrave, W.D. and Timmons, J.A. (1992) *Venture Capital at the Crossroads*, Harvard Business School Press, Boston.
Cary, L. (1993) *The Venture Capital Report Guide to Venture Capital in Europe*, Pitman, London, 6th edition.
CBI (1983) *Smaller Firms in the Economy*, Confederation of British Industry, London.
Conlin, E. (1989) 'Adventure capital', *INC Magazine*, September, pp. 32–48.
Deakins, D. and Hussain, G. (1991) *Risk Assessment By Bank Managers*, Department of Financial Services, Birmingham Polytechnic Business School, Birmingham.
Department of Trade and Industry (1992) *Companies in 1990–91*, HMSO, London.
Dixon, R. (1989) 'Venture capitalists and investment appraisal', *National Westminster Bank Quarterly Review*, November, pp. 2–21.
Economist (1991) 'Venture capital: plenty to gain?' 7 December, p. 126.
Ennew, C. and Binks, M.B. (1993) 'Financing entrepreneurship in recession: does the banking relationship constrain performance?', paper presented at the 13th Babson Entrepreneurship Research Conference, University of Houston.
Fiet, J.O. (1991) 'Network reliance by venture capital firms and business angels: an empirical and theoretical test', in Churchill, N.C., Bygrave, W.D., Covin, J.G., Sexton, D.L., Slevin, D.L., Vesper, K.H. and W.E. Wetzel, W.E. (eds) *Frontiers of Entrepreneurship Research 1991*, Babson College, Babson Park: MA, pp. 445–55.
Financial Times (1993a) 'Business adjusts to tighter credit', 27 February, p. 5
Financial Times (1993b) 'Businesses "are not ready for recovery"', 25 January, p. 7
Forth, E. (1992) 'Bridging the capital gap for small businesses', *Venture Capital Report*, March, mid-page article.
Freear, J. and Wetzel, W. (1988) 'Equity financing for new technology-based firms', in Kirchhoff, B.A., Long, W.A., McMullen, W.E., Vesper, K.H. and Wetzel, W.E. (eds) *Frontiers of Entrepreneurship Research 1988*, Babson College, Babson Park: MA, pp. 347–67.
Gapper, J. (1993a) 'Lessons of the 80s spark policy review', *Financial Times*, 27 February, p. 5.
Gapper, J. (1993b) 'The equation that did not add up', *Financial Times*, 2 February 1993, p. 15.
Gaston, R.J. (1989a) 'The scale of informal capital markets', *Small Business Economics*, 1, 223–30.
Gaston, R.J. (1989b) *Finding Private Venture Capital For Your Firm: A Complete Guide*, Wiley, New York.
Gaston, R.J. and Bell. S.E. (1986) *Informal Risk Capital in the Sunbelt Region*, US Small Business Administration, Office of Advocacy, Washington DC.
Gaston, R.J. and Bell, S.E. (1988) *The Informal Supply of Capital*, US Small Business Administration, Office of Advocacy, Washington DC.

Goodman, J.P. and Allen, K.R. (1992) 'The credit crunch: are Federal policies putting entrepreneurial firms on a debt diet?', paper presented at the 12th Babson Entrepreneurship Research Conference, INSEAD, Fontainebleau.

Government of Ontario (1990) *The State of Small Business: 1990 Annual Report on Small Business in Ontario*, Ministry of Industry, Trade and Technology, Toronto.

Haar, N.E., Starr, J. and MacMillan, I.C. (1988) 'Informal risk capital investors: investment patterns on the East Coast of the USA', *Journal of Business Venturing*, 3, 11–29.

Harrison, R.T. and Mason, C.M. (1991) 'Informal investor networks: a case study from the United Kingdom', *Entrepreneurship and Regional Development*, 3, 269–79.

Harrison, R.T. and Mason, C.M. (1992a) 'International perspectives on the supply of informal venture capital', *Journal of Business Venturing*, 7, 459–75.

Harrison, R.T. and Mason, C.M. (1992b) 'The roles of investors in entrepreneurial companies: a comparison of informal investors and venture capitalists', *Venture Finance Working Paper No. 5*, Urban Policy Research Unit: University of Southampton, Southampton.

Institute of Directors (1992) *Small Firms in the UK Economy: A Business Leader's View*, IoD, London.

Mason, C.M. and Harrison, R.T. (1991) 'Venture capital, the equity gap and the north–south divide in the UK', in Green, M. (ed.) *Venture Capital: International Comparisons*, Routledge, London, pp. 202–47.

Mason, C.M. and Harrison, R.T. (1992) 'The supply of equity finance in the UK: a strategy for closing the equity gap', *Entrepreneurship and Regional Development*, 4, pp. 357–80.

Mason, C.M. and Harrison, R.T. (1993a) 'Informal risk capital: a review of US and UK evidence', in Atkins, R., Chell, E. and Mason, C. (eds) *New Directions in Small Business Research*, Avebury, Aldershot, pp. 155–76.

Mason, C.M. and Harrison, R.T. (1993b) 'Strategies for expanding the informal venture capital market', *International Small Business Journal*, 11 (4), 23–38.

Mason, C.M. and Harrison, R.T. (1993c) 'Promoting informal venture capital: an evaluation of a British initiative', paper presented at the 13th Babson Entrepreneurship Research Conference, University of Houston.

Mason, C.M. and Harrison, R.T. (1993d) *Interim Review of Five Informal Investment Demonstration Projects*, Sheffield, Department of Trade and Industry: Small Firms Policy Branch.

Mason, C., Harrison, R. and Chaloner, J. (1991a) 'Informal risk capital in the United Kingdom: a study of investor characteristics, investment preferences and decision-making', *Venture Finance Research Project Working Paper No. 2*: Urban Policy Research Unit, University of Southampton, Southampton.

Mason, C.M., Harrison, R.T. and Chaloner, J. (1991b) *The Operation and Effectiveness of LINC. Part 1: Survey of Investors*: Urban Policy Research Unit, University of Southampton, Southampton.

Mason, C.M., Harrison, J. and Harrison, R.T. (1988) *Closing the Equity Gap? An Assessment of the Business Expansion Scheme*, Small Business Research Trust, London.

110   Colin Mason and Richard Harrison

McMeekin, D. (1991) 'Finance for enterprise: closing the equity gap', paper presented at the 14th Small Firms Policy and Research Conference, Lancashire Enterprises Ltd/Manchester Business School.

Midland Bank (1992) *The Changing Financial Requirements of Smaller Companies*, Midland Bank Business Economics Unit, London.

Miller, B. (1992) 'Angelic mission to make money', *CA Magazine*, November, 14–18.

Murray, G. (1990) *Change and maturity in the UK venture capital industry 1990–95*, Warwick Business School, Coventry.

Murray, G. (1993) 'Venture capital', paper presented at a CBI Workshop on Finance for SMEs.

Myers, D.D. and Moline, M.D. (1988) 'Network and participation interest of Missouri informal investors', paper presented at the 8th Babson Entrepreneurship Research Conference, Calgary.

NEDC (1986) *External Capital for Small Firms*, National Economic Development Office, London.

Neiswander, D.K. (1985) 'Informal seed stage investors', in Hornaday, J.A., Shils, E.B., Timmons, J.A. and Vesper, K.H. (eds) *Frontiers of Entrepreneurship Research 1985*, Babson College, Babson Park: MA, pp. 142–54.

Oates, D. (1992) 'Visions of angels', *The Director*, June, 35–8.

Outram, R. (1992) 'Nothing ventured, nothing gained?', *Accountancy Age*, February, 8–12

Peek, J. and Rosengren, E.S. (1992) 'The capital crunch in New England', *New England Economic Review*, May/June, pp. 21–31.

Plender, J. (1993) 'Caught in a double bind', *Financial Times*, 6/7 March, p. 8.

Postma, P.D. and Sullivan, M.K. (1990) *Informal Risk Capital in the Knoxville Region*, Centre of Excellence for New Venture Analysis, College of Business Administration, The University of Tennessee, Knoxville, TN.

Riding, A.L. and Short, D.M. (1987a) 'Some investor and entrepreneur perspectives on the informal market for risk capital', *Journal of Small Business and Entrepreneurship*, 5 (2), 19–30.

Riding, A.L. and Short, D.M. (1987b) 'On the estimation of the investment potential of informal investors: a capture–recapture approach', *Journal of Small Business and Entrepreneurship*, 5 (4), 26–40.

Short, D.M. and Riding, A.L. (1989) 'Informal investors in the Ottawa–Carleton region: experiences and expectations', *Entrepreneurship and Regional Development*, 1, 99–112.

Small Business Research Trust (1991) 'Small business finance', *NatWest Quarterly Survey of Small Business in Britain*, 7 (4), 19–21.

Spragins, E.E. (1991) 'Heaven sent', *INC Magazine*, February, 85–7.

Sullivan, M.K. (1991) 'Entrepreneurs and informal investors: are there distinguishing characteristics?' in Churchill, N.C., Bygrave, W.D., Covin, J.G., Sexton, D.L., Slevin, D.L., Vesper, K.H. and Wetzel, W.E. (eds) *Frontiers of Entrepreneurship Research 1991*, Babson College, Babson Park: MA, pp. 456–68.

Timmons, J.A. (1990) *Planning and Financing the New Venture*, Dover: MA, Brick House Publishing Co.

Tirbutt, E. (1993) 'Rough guide to the risk takers', *Accountancy Age*, February, 38–43.
Tymes, E.R. and Krasner, O.J. (1983) 'Informal risk capital in California', in Hornaday, J.A., Timmons, J.A. and Vesper, K.H. (eds) *Frontiers of Entrepreneurship Research 1983*, Babson College, Babson Park: MA, pp. 347–68.
Vyakarnam, S. and Jacobs, R. (1991) 'How bank managers construe high technology entrepreneurs', paper presented at the National Small Firms Policy and Research Conference, Lancashire Enterprises Ltd/Manchester Business School.
Wetzel, W.E. (1981) 'Informal risk capital in New England', in Vesper, K.H. (ed.) *Frontiers of Entrepreneurship Research 1981*, Babson College, Babson Park: MA, pp. 217–45.
Wetzel, W.E. (1983) 'Angels and informal risk capital', *Sloan Management Review*, 24, summer, pp. 23–34.
Wetzel, W.E. (1986a) 'Entrepreneurs, angels and economic renaissance', in Hisrich, R.D. (ed.) *Entrepreneurship, Intrapreneurship and Venture Capital*, Lexington Books, Lexington: MA, pp. 119–39.
Wetzel, W.E. (1986b) 'Informal risk capital: knowns and unknowns', in Sexton, D.L. and Smilor, R.W. (eds) *The Art and Science of Entrepreneurship*, Ballinger, Cambridge: MA, pp. 85–108.
Wetzel, W.E. jnr (1987) 'The informal risk capital market: aspects of scale and efficiency', in Churchill, N.C., Hornaday, J.A., Kirchhoff, B.A., Krasner, O.J. and Vesper, K.H. (eds) *Frontiers of Entrepreneurship Research 1987*, Babson College, Babson Park: MA, pp. 412–28.
Wetzel, W.E. and Seymour, C. (1981) *Informal Risk Capital in New England*, US Small Business Administration, Office of Advocacy, Washington DC.
Woodcock, C. (1991) 'Angels are waiting in the wings', *Guardian*, 18 November.
Woodcock, C. (1992a) 'Angels step in where bankers fear to tread', *Guardian*, 12 October, p. 12
Woodcock, C. (1992b) 'A more equitable way to capitalise on investment', *Guardian*, 28 September, p.12

4 Financial constraints to the growth and development of small high-technology firms

Barry Moore

INTRODUCTION

There is now a significant body of research on the financial problems of small firms and the importance of the availability and cost of finance to firms seeking to expand and develop, ACOST (1990), Hutchinson and McKillop (1992). The recent DTI (1991) report is the result of but one of a number of government enquiries (MacMillan, Radcliffe, Bolton and Wilson committees) appraising the severity of problems experienced by small firms in accessing finance. Recent research in the United States also links the different stages in the growth of small high-technology firms to the parallel development of their financial needs and the problems they face in meeting them (Roberts 1991). Problems in accessing external sources of finance have been identified during the start up phase, particularly finance from banks and it is clear that the nature and source of external finance for the evolving small firm changes through the life cycle of development. Storey and Strange (1992) confirm that personal savings remain the most important source of start up funding but internally generated profits and external funding from banks assume much greater significance for the more mature small firm (Oakey *et al*. 1990, Roberts 1991). It is also recognized, ACOST (1990) that because the evolution of the firm is characterized by discontinuities, reflecting for example shifts in market opportunities or technological developments, injections of external funds will be required if the firm is to successfully adapt to changing conditions and maintain its competitive position. Such disjunctures will also often raise important challenges for the management and organization of the small firm. Although these problems and potential obstacles to growth confront all small firms to a greater or lesser degree, arguably they are more severe for the small high-technology firm where the product/process is often untested in the market, where rapidly developing new technology makes existing technology obsolete very quickly and where there is often a relentless process of learning and developing in circumstances of considerable uncertainty (Slatter 1992, Van Glinow and Mohrman 1990). That small high-technology firms might face relatively more severe problems in accessing finance, particularly debt finance, is also supported by theoretical work which emphasizes information asymmetries between financial institutions and the small firm (Binks, Ennew, Reed 1992). These information problems affect the willingness of banks to enter into contracts to supply debt and in these circumstances collateral and personal finance become critical for start up and subsequent expansion. A study by Hunsdiek and Albach (1988) of new technology-based firms in Germany revealed that one-third of all the companies founded started with no debt capital at all and one-fifth of all founders said that the banks wanted too much collateral.

This chapter examines focuses on financial constraints facing small high-technology companies and draws on data derived from two important company surveys recently carried out in the Cambridge University Small Business Research Centre (SBRC).[1] A distinctive feature of part of the analysis facilitated by this database, is a comparison of small high-technology firms with small 'conventional' firms to establish whether the former do indeed face more severe financial constraints because of the relatively high degree of risk inherent in a new technology-based enterprise.

The first part of the analysis concentrates on financial constraints in the start up and early stages of the new high-technology business. The second section examines financial constraints at a somewhat later stage in development when the firm has achieved a degree of maturity and is perhaps seeking to diversify and expand. The third part of the chapter focuses on financial constraints in the period subsequent to start up and tests whether small high-technology firms are more likely to face financial constraints in securing their business objectives than 'conventional' small firms. The next section tests whether small high-technology firms are more likely than 'conventional' firms to face financial constraints in introducing new technology. The final section of the chapter analyses the relative importance of different sources of external finance used by small high-technology companies.

FINANCIAL CONSTRAINTS IN THE START UP PHASE[2]

It is widely recognized that all new business start ups are inherently risky and failure rates are very high in the first few years after birth.

Not surprisingly therefore, obtaining finance is a problem faced by a significant proportion of small firms starting up. The DTI (1991) study reported that 20 per cent of start-up firms (i.e. firms established in the last two years) experienced problems in raising finance, although this proportion was virtually the same for firms of all ages. Moreover established businesses seeking to introduce product and process innovations were especially likely to experience constraints in raising finance by comparison with other businesses. Thus there are grounds for believing that start-up firms in the high-technology sector are not only dependent on risk-bearing capital but are more likely to face problems in raising their initial start-up capital than 'conventional' start ups. In the period from conception of the product/service to formal start up, high-technology companies are very often continuing to solve product and development issues, modifying the product and testing prototypes with pilot customers. The product itself may often be innovative to the point where little market testing has been possible and the more innovative the product the more likely it will require a complex interactive process of feedback and redesign before it is suitable for market launch. The technology may be insufficiently understood by potential investors and the regulatory framework, particularly with new medical products, may add to the uncertainty surrounding their potential use and market significance. Thus, Roberts and Hauptman (1987) note that regulatory compliance in the biomedical industry may not only delay product introduction and company launch but also raise the threshold financial capital needed to start the business by comparison with start ups in the 'conventional' sector. High-technology firms which have their origin in university or other research establishments also face difficulties involving intellectual property rights; and academic high-technology spin-off companies may often be launched with inadequate capitalization because of a reluctance of inventors to dilute their equity stake (Roberts 1991). At the same time high-technology companies whose founders have a primarily research or technical background, may be ill-prepared to meet the demanding requirements of banks and other potential investors, although these problems may be significantly less severe where the founder has a demonstrable record of success in the industry. External finance may also be difficult to acquire in circumstances where capital equipment is very specific to the product itself and of little value should the firm fail. Moreover, for many high technology firms the resources are intangible and vested very much in the founder. What Bullock (1983) has termed 'soft starts', involving consultancy or contract work on a bespoke basis, with minimal R&D and modest capital requirements, perhaps provides further evidence on finance constraints at start up for many high-technology firms. 'Hard-starts' with their higher downside risks and greater resource commitments are less likely to find favour with conservative bankers even though the ultimate returns may be higher than a 'soft start'.

These considerations suggest that a variety of factors including *inter alia* the innovativeness of the product, maturity of the market, the specificity of capital inputs required, the technology, the often intangible nature of the 'capital' base, the previous experience of the founders and the attitudes, practices and imperfections in the capital market, combine to a greater or lesser degree to present constraints and difficulties in securing start-up finance for the high-technology firm. Expanding on this, our research programme distinguished different types of start-up companies according to the extent of their financial requirements. In part the distinction is determined by the type of technology and its stage of development and in part by whether market entry is of the 'hard' or 'soft start' variety.

Three types of companies can be distinguished at the start up phase: (a) companies developing products requiring high front-end development costs; (b) companies where lead times from concept to market launch are short, front-end development costs are low and as a consequence cash flow generation occurs relatively early in the company life cycle; (c) companies where market entry and product development occurs over an extended period of time often through consultancy, contract R&D in niche markets. Companies in category (a) would normally require significant seed capital funding by comparison with companies in categories (b) and (c). Many biotechnology companies tend to fall into category (a) as they struggle to transform their research into marketable products. The need for finance to develop prototypes and market test product during the life cycle stage from concept to market launch is great. For other technologies where development is less turbulent stage finance needs at start up may be less great and more easily satisfied.

The relative importance of finance as a problem at start up

Although it is widely recognized that finance is an important problem at start up for small high-technology companies, there is relatively little research which systematically sets the finance issue in the context of the range of business development problems that companies face at this early stage in their life cycle. Hunsdiek and Alback

(1988) indicate the preponderance of financial problems relative to other problems in their study of German new technology-based start ups. Financing was most frequently referred to (27 per cent) followed by complexity of regulation (20 per cent). Slatter (1992) drawing on case study material lists, a wide range of start-up problems that potentially create cash-flow problems for high-technology start ups, but his research is primarily concerned with strategic and organizational practices in managing growth. Other research in the UK generally fails to focus on small high-technology firms at start up and the *relative* importance of finance as a constraint.

In the face-to-face survey of eighty-nine high-technology companies a range of questions was asked relating to business problems faced when starting up. Some forty questions were asked covering four broad areas of business activity – product development and market intelligence, obtaining finance and managing cash flow, management and organization, selecting and developing a location and site. Table 4.1 shows within each broad category of business activity the number of times a particular business activity or requirement was mentioned as presenting a major problem and the percentage frequency.

Obtaining and managing finance and product development and market intelligence emerge as the most important sources of major problems for small high-technology firms. The high ranking of finance is not surprising given the uncertainty and risk associated with any start up, particularly a high-technology start up. That product development should also rank highly as a source of major problems is also not surprising and is supported by research in the US (Roberts 1991). Product development by sophisticated entrepreneurs with a high degree of technical competence is very often technology-led rather than market-led. Thus products may be technically advanced, but with little customer appeal. Moreover, the technically gifted entrepreneur often over responds to specific customer requirements by redesigning the product before allowing sufficient time for his intended standard product to sell. Roberts (1991) also found that lack

Table 4.1 Number and percentage frequency with which high-technology companies claimed to be facing a major problem at start up

Business activity	Number	%
Product development, market intelligence	155	16
Obtaining and managing finance	111	16
Management and organization	146	11
Site selection and location	42	8

Table 4.2 Financial activities presenting small high-technology companies with a major problem at start up and early stage development

	Number	*%*
Managing cash flow	21	24
Obtaining debt financing	16	18
Securing adequate operating capital	21	24
Obtaining equity financing	14	16
Credit control and invoicing	9	10
Managing capital	13	15
Establishing a banking relationship	7	8
Developing an accounting/control system	10	11

Source: Interview survey

of market-orientation at the outset not only presented problems at start up, but also resulted in only limited subsequent correction with formal market activities still absent from high-technology firms several years later. The high ranking of management and organizational problems may also reflect the tendency of founders of technology-based start ups to devote too much time to solving technical problems and too little time to the management and organization of the business.

Within the range of issues raised under financial problems, managing cash flow was cited most often as presenting major problems, whilst debt financing, operating capital and equity finance presented problems for about one quarter of the sample, Table 4.2.

Business spin-offs generally experienced less severe problems with finance than other high-technology companies starting up (see Table 4.3). In part, this reflects their greater reliance on self-financing but it is also bound up with the greater credibility and experience that they are likely to have gained in the industry by comparison with founders coming out of HEIs or completely new start ups. Even so, managing cash flow and securing adequate finance was a major problem for about a quarter of companies including business spin-offs.

Main sources of funding at start up

The most important source of start-up funding is personal or self-finance and this is particularly so for business spin-offs and completely new companies (see Table 4.4).

By contrast, outside funding is important for the majority (58 per cent) of spin-offs from universities and government research

Table 4.3 Financial activities presenting small high-technology companies with a major problem by origin of start up (%)

Activities	Business spin-off	Research spin-off	Completely new	Other	Total
Obtaining equity financing	–	17	24	12	16
Obtaining debt financing	11	8	24	18	18
Establishing a banking relationship	5	–	12	5	8
Developing an accounting/control system	5	8	17	6	11
Managing capital	16	8	22	–	15
Managing cash flow	26	8	34	6	24
Credit control and invoicing	11	–	15	6	10
Securing adequate operating capital	16	25	32	12	24
Total number of companies	19	12	41	17	89

Table 4.4 Main sources of start up finance by origin of small high-technology firm (%)

Source of start up finance	Business spin-off	Research HEI	Completely new	Other	Total
Overwhelmingly self-finance	53	17	50	41	44
Overwhelmingly outside funding	37	58	23	29	32
Balanced mix of self and outside finance	11	25	28	29	24
	100	100	100	100	100
Number of companies	19	12	40	17	88

Source: Interview survey

establishments and this may be because very often the academic researcher seeks business partners from outside, or it may merely reflect the relatively low income of potential academic entrepreneurs and the difficulties of accumulating savings. There is also evidence that in a limited number of cases the university has acted as a guarantor for a start up which would help in raising external finance.

The follow up survey to the initial interview programme shows that 52 per cent of start-up finance came from founders savings, family and friends. A recent survey by Stoy Hayward (1992) shows that for small firms in general, 51 per cent of money invested in independent new ventures was personal finance. Storey and Strange (1992) also

Table 4.5 Sources of finance for high-technology start ups (mean %)

Source	%
Founders savings	49
Bank loans	7
Money from family and friends	9
Money from government agencies	9
Venture capital	10
University endowments	6
Strategic partners	6
Total	100

Source: Follow-up survey (No. of companies 42)

find that for small firms generally, personal finance made up about half of all sources of finance mentioned. However, loans or overdrafts from banks were of next most importance in the Storey and Strange study, providing about one quarter of finance used, compared with only 7 per cent for high-technology firms (see Table 4.5). The Stoy Hayward (1992) results also confirm the much smaller share of financial support provided by the banks to the high-technology small firms compared with small firms generally (40 per cent).

This confirms the view that banks are generally more reluctant to support the start up of high-technology new ventures by comparison with start ups in the 'conventional' sector. It is interesting to note that similar data produced by Roberts (1991) showed that in his sample of 154 US high-technology companies, no initial support was provided by the commercial banks, although bank credit was available very early on in the life of the company. Oakey (1984) also provides evidence on the relative unimportance of bank finance at the start-up stage accounting for 6–7 per cent of total sources of finance.

Venture capital provided some 10 per cent of finance for start up but funding from this source is notoriously volatile, being dependent on changing financial conditions, the business cycle and the technology cycle. Currently (1992) venture capital is providing only a very limited contribution to the funding of high-technology starts and the climate for venture capital activity is likely to remain inclement for some time (Murray 1992).

Although bank finance ultimately provided a relatively small proportion of start-up finance for high-technology companies, over two-thirds of companies *considered* the banks as a potential source

Table 4.6 Sources of finance considered by start up high-technology company (% of companies)

	%
Founders savings	69
Bank loans	69
Money from family/friends	12
Loans from government agencies/relatives	48
Venture capital	45
University endowments	5
Strategic partners	29

Source: Follow-up survey (No. of companies 42)

(Table 4.6). Similarly venture capital and loans from government, although considered by nearly half the companies, failed to materialize for the majority of these companies.

It is clear therefore that although companies did face problems in attracting finance at start up (58 per cent of the forty-three companies) such problems were very often overcome one way or another, with founders' savings playing a major role. Finally it is perhaps worthy of note that 60 per cent of the eighty-nine companies interviewed rated the government's role in initial funding as 'not good at all'.

Some implications of difficulties in obtaining start up finance

Difficulties in obtaining start-up capital and the adequacy of initial capitalization is reflected in the subsequent performance of the high-technology firm (see Table 4.7).

However, one must be careful not to rule out the possibility that the absence or less frequent experiences of difficulty by 'stars' in obtaining start-up capital may merely reflect the influence of other factors indicative of their potential success at start up. Recent research by Hall and Young (1991) and Hall (1992) also identifies

Table 4.7 Difficulties in obtaining start up capital and subsequent growth (% employment change 1985–92)

Companies	Flop	Below average	Above average	Star	Total
% facing difficulties	50	33	26	–	–
Number of companies	6	21	19	9	55

Source: Interview survey
Notes: Flop: bottom decile, Star: top decile, Below/Above average: 40 per cent each

Table 4.8 Original plans for growth and source of initial finance (%)

Original plans	Self-finance	Source of initial funding mainly Self and external	External	Total
No plan	10	1	7	7
Start and stay small	18	10	19	16
Grow slowly	49	43	22	40
Grow fast	23	48	52	38
	100	100	100	100
Number of companies	39	22	27	88

under-capitalization as the most important reason for small firm insolvency and Roberts (1991) finds a close positive association between the performance of high-technology companies and their initial financing.

Finally companies where external finance is of overwhelming importance tend to be more growth orientated at start up than companies relying primarily on self-finance (see Table 4.8).

FINANCIAL CONSTRAINTS IN EARLY GROWTH AND SUBSEQUENT MATURITY

This section analyses the role of financial constraints in limiting the achievement of business objectives as the small high-technology firm matures and develops. A distinctive feature of the analytical approach adopted is the comparison of small high-technology firms with firms in the 'conventional' sector. Such a comparison permits the identification of financial constraints which differentiate between small high-technology companies and 'conventional' small firms.

The relative importance of financial constraints by sector

A variety of factors can limit the ability of a business to meet its business objectives. The ACOST (1990) Report focused on markets, resources, technology and management and organization as the main areas where barriers to growth might be found. The SBRC survey deployed a similar classification to that adopted by ACOST but at more disaggregated level and identified eleven broad areas of potential constraint, including the availability and cost of finance for expansion and the availability and cost of overdraft finance. Surveyed firms were requested to rank each of eleven factors from 0–9 with 0 being completely unimportant and 9 being highly important, which

Table 4.9 Constraints on the ability to meet business objectives for high-technology and 'conventional' sectors (mean of rankings)

Constraint	Conventional			High-technology		
	Mfg	Ser	Tot	Mfg	Ser	Tot
Finance						
Availability & cost of finance for expansion	5.4	4.9	5.2	5.1	5.5	5.4
Availability & cost of overdraft finance	5.4	4.9	5.2	5.0	5.3	5.2
Labour						
Skilled labour	3.9	3.4	3.7	3.6	4.0	3.9
Management						
Management skills	4.2	4.1	4.2	4.3	4.6	4.5
Marketing & sales skills	4.3	4.3	4.3	5.0	5.4	5.3
Technology						
Acquisition of technology	2.7	2.5	2.6	3.5	2.0	2.5
Difficulties in implementing new technology	2.5	2.3	2.4	3.5	2.0	2.5
Premises						
Availability of suitable premises and site	2.4	2.3	2.4	2.1	2.3	2.2
Markets						
Access to overseas markets	2.4	1.7	2.1	3.1	2.0	2.3
Overall growth of market demand	5.4	5.0	5.2	5.3	4.7	4.9
Increasing competition	5.0	4.5	4.8	4.4	4.2	4.3

Source: SBRC Survey (1992)
Note: Over 1500 companies in the conventional sector and nearly 300 in high-technology

may have acted as a significant limitation to meet their business objectives over the past three years. Table 4.9 shows the means of the rankings for high-technology firms and 'conventional' firms for each of the factors.

The availability and cost of finance for expansion represents the most important constraint facing high-technology firms and this is also true for 'conventional' firms, although it is marginally more severe for the former group of firms. There is also an indication that small high-technology service sector firms are more severely constrained than both 'conventional' service sector firms and high-technology manufacturing firms. Moreover high-technology manufacturing firms would appear to face slightly less severe financial constraints for expansion than 'conventional' manufacturing small firms. Limitations with respect to marketing and sales skills ranked second in importance on average for high-technology firms and the mean rank of 5.3 is very much higher than that for 'conventional'

Table 4.10 Constraints on the ability to meet business objectives for small firms in high-technology and 'conventional' sectors (percentage)

Constraint	Conventional		High-technology	
	Completely unimportant	*Highly important*	*Completely unimportant*	*Highly important*
Finance				
Availability and cost of finance for expansion	20	30	20	33
Availability and cost of overdraft finance	21	27	21	30
Labour				
Skilled labour	27	7	25	10
Management				
Management skills	20	8	15	8
Marketing and sales skills	19	9	11	13
Technology				
Acquisition of technology	38	3	38	2
Difficulties in implementing new technology	41	3	37	2
Premises				
Availability of suitable premises and site	50	6	49	5
Markets				
Access to overseas markets	51	3	49	4
Overall growth of market demand	19	23	18	18
Increasing competition	14	10	16	7

Source: SBRC Survey (1992)

firms (3.5). Again this constraint is somewhat more severe for high-technology service companies than for high-technology manufacturing companies. Constraints imposed by the availability and cost of overdraft finance are similar for high-technology and 'conventional' firms and ranked third in overall importance for the firm in trying to secure its objectives. Limitations arising from market conditions are ranked next, although it is the small firms in the conventional sector that view this factor as more severe than small high-technology firms. 'Conventional' manufacturing firms seem particularly under pressure from this constraint with a ranking of 5.4 for overall growth of market demand, equal to that of the constraint posed by finance. Perhaps surprisingly, difficulty in accessing overseas markets was not ranked highly as a constraint for either high-technology or conventional small firms despite the well known problems of going international (Slatter 1992).

It is also instructive to show the proportion of small companies in the two sectors claiming that different constraints were highly important (rank 9) or completely unimportant (rank 0) (see Table 4.10).

One striking feature of the results presented in Table 4.10 is the similarity of the proportion of conventional and high-technology firms claiming that the same constraints are highly important. Thus, although the availability of finance for expansion was a highly important constraint for a significant proportion of firms, there was little difference between high-technology small firms (33 per cent) and conventional small firms (30 per cent).

Finance constraints and the age of company

In the period following the initial start-up phase the small firm will be relying on a combination of internally generated funds and external funds for working capital and finance for expansion. In this period it is very likely that the small firm will have tied up a significant proportion of its initial capital in fixed assets and will not as yet have built up a significant flow of profits. Credit from suppliers may also be difficult to secure in the early years, resulting in further strains on working capital. As the small firm matures and establishes profitability, pressures on working capital may initially ease; but with expansion into new markets, further R&D and increased investment and output, pressures on working capital are likely to persist and indeed constrain the growing firm. Moreover, growth of the company will require additional finance to support expansion, which again must be met partly out of retained profits and partly from external funds. Table 4.11 shows how the availability and cost of finance changes in severity, limiting the achievement of the firms objectives as it matures through the life cycle of growth and development. For both constraints there is a suggestion of a very slow decline in the importance of financial constraints as the firm ages with the peak values for the youngest firms. Moreover, the mean rank does not decline smoothly with age of firm. Thus, with respect to the availability and cost of finance for expansion, there appears to be a four-year peaking at age 5, 9, 13, 17 and a further peak of 20. A similar but not identical pattern exists for overdraft finance. Thus although in broad terms finance as a constraint diminishes as the company matures, it does so at a relatively slow rate.

Figure 4.1 illustrates the steadily diminishing importance of finance as a limitation on the achievement of the small firm's objectives by focusing on the proportion of all small firms in the sample claiming

Table 4.11 The severity of limitations arising from the availability and cost of finance by age of firm (means of ranked data)

Age	No. of firms	Availability and cost of finance for expansion	Availability and cost of overdraft finance
2	20	6.2	6.9
3	53	6.6	5.7
4	91	5.6	5.8
5	110	6.1	5.4
6	103	5.5	5.5
7	97	5.4	5.2
8	95	5.9	5.5
9	77	6.0	5.7
10	87	5.3	5.0
11	85	5.2	5.1
12	69	5.3	5.1
13	67	6.0	6.1
14	49	5.9	5.9
15	34	4.5	4.7
16	42	5.0	5.4
17	36	5.9	5.6
18	34	5.1	5.3
19	39	4.3	4.4
20	41	5.0	4.8
21–50	348	4.6	4.8
50+	210	4.4	4.5

Source: SBRC Survey (1992)

finance limitations to be highly important (rank 9) or unimportant (rank 0). At a very early stage of development some 40 per cent of firms claim that the availability and cost of finance is a highly important constraint which limits their ability to achieve their objectives. As the firm matures this constraint is of diminishing importance, and at age 20 less than one-third of firms claim that it is highly important.

Figure 4.2 compares the proportion of companies claiming that financial constraints are highly important in the high-technology and 'conventional' sectors. Although there is some indication that high-technology companies are more likely to experience highly important financial constraints than small conventional firms, the evidence is by no means conclusive.

Finance constraints and growth

It was argued previously that there is evidence to suggest that small fast growing businesses face difficulty in accessing finance to support

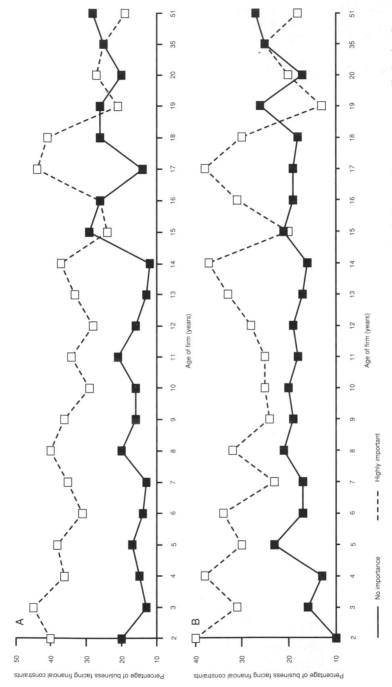

Figure 4.1 Availability and cost of (a) finance for expansion and (b) cost of overdraft finance: importance as a limitation in meeting business objectives

their growth and development (Storey, Watson and Wynarczyk 1989). Local branches of clearing banks are reluctant to fully support the fast growing firm, particularly where insufficient collateral is available. Given the nature of the typical small high-technology firm one might expect such difficulties to be even greater by comparison with small fast growing firms generally. Table 4.12 provides some support for this view. In the manufacturing sector the fastest growing firms (stars) are more likely to be limited by financial constraints than are firms growing more slowly. Moreover the evidence supports the view that high-technology firms are more likely to be constrained than conventional firms. For the service sector the evidence is less conclusive. For high-technology firms constraints are more likely to be associated with fast growth when growth is measured by sales performance and less likely when measured by employment change. For small conventional firms in the service sector the probability of facing a financial constraint regarded as highly important *falls* as growth increases on both measures of firm growth.

Regression analysis

This section uses a regression analysis to test whether small firms in the high-technology sector have a higher probability of facing problems with the availability and cost of finance in achieving their business objectives than do conventional small firms. The statistical analysis also permits the identification of characteristics of the small firm other than sector which are associated with finance problems.

A probit analysis was carried out for the two dependent variables, namely the availability and cost of finance for expansion and the availability and cost of overdraft finance. For each of these variables probit regressions were undertaken with the dependent variable taking a value according to the ranking given to the importance of the constraint. Four sets of rankings were tested within the range 0 (completely unimportant) to 9 (highly important) and included rank 9, ranks 8 and 9 together, rank 0 and ranks 0 and 1 together. Thus, for rank 9, for example, the regression would test for finance being highly important, relative to all other ranks, against each of the independent variables.

The independent variables included in the regression covered firstly the activity of the firm classified into four groups – high-technology manufacturing and services and conventional manufacturing and services. This activity classification is designed specifically to test for the relative importance of access and cost of finance in

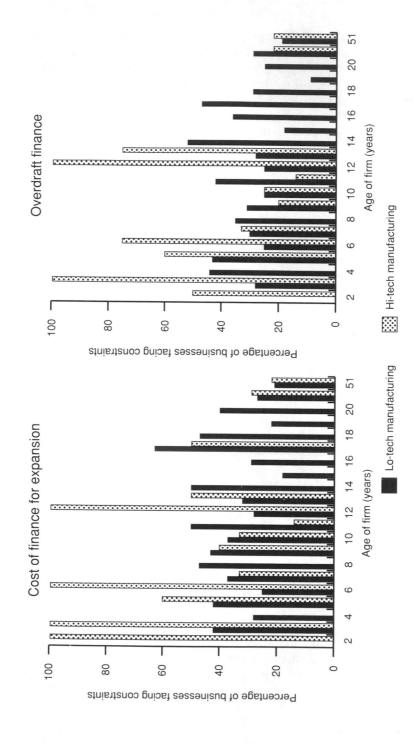

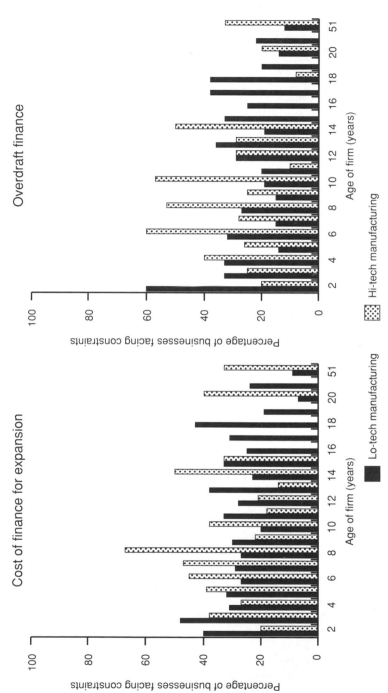

Figure 4.2 Availability and cost of finance for expansion and overdraft finance: per cent of high-technology and conventional firms saying highly important

130   *Barry Moore*

Table 4.12 The importance of the availability and cost of finance and the growth of the small firm in the high-technology and 'conventional' sectors (% indicating highly important)

% Sales growth 1987/90	Availability and cost of finance for expansion				Availability and cost of overdraft finance			
	High-tech		Conventional		High-tech		Conventional	
Manufacturing								
Flops	33	(36)	26	(31)	22	(27)	29	(34)
Below average	33	(27)	28	(26)	30	(23)	26	(24)
Above average	22	(36)	34	(36)	20	(28)	29	(29)
Stars	57	(43)	37	(34)	50	(50)	33	(34)
Services								
Flops	24	(50)	26	(30)	18	(36)	25	(31)
Below average	25	(29)	17	(21)	25	(27)	16	(18)
Above average	31	(25)	29	(30)	27	(24)	25	(25)
Stars	46	(38)	21	(22)	38	(33)	13	(18)

Source: SBRC Survey (1992)
Note: Figures in parentheses refer to employment growth 1987–90.

constraining the high-technology firms. Secondly, the age of the firm was included to test for life-cycle effects, with the expectation that more mature small firms find raising finance less of a problem than younger firms. Thirdly the origin of the firm is included, distinguishing spin-off from existing business, management buy-out, merger with or purchase of existing firm(s) and completely new start ups. Whether the business was a consequence of actual or potential unemployment was also included. A priori it might be expected that where credibility and experience has been gained as a result of previous employment in an existing business there should be a below average probability of experiencing highly important constraints. Fourthly, the rate of growth of the small firm was included as measured by both sales and employment growth between 1987 and 1990. Fifthly, the profitability of the firm was included (pre-tax profits as a share of turnover) with the expectation that more profitable firms should experience less severe limitations in raising finance. The size of firm, the extent of its involvement in export markets (exports as a share of turnover) and whether the firm had introduced major innovations in the past five years were also included in the regressions. Finally the analysis tested for a North/South effect influencing the difficulties with finance for small firms.

The results of the regression analysis are shown in Tables 4.13 and 4.14.

Table 4.13 shows that the small firms in the high-technology service sector have an above average probability of finance for expansion

Table 4.13 Regressions results: dependent variable – availability and cost of finance for expansion

Independent Variable [1]	Ranking of importance — Unimportant — 0 and 1 Coeff.	0 and 1 t-ratio	Unimportant 0 Coeff.	0 t-ratio	8 and 9 Coeff.	8 and 9 t-ratio	Highly important 9 Coeff.	9 t-ratio
Activity [1]								
High-tech man.	0.1E1	0.6	-0.2E1	-0.1	0.2	1.5	0.2	1.2
High-tech serv	-0.1	-1.0	-0.1	-1.2	0.2	1.7	0.1	1.0
Conv. man	-0.2	-2.2	-0.2	-2.2	0.1	1.8	0.2	2.3
Life cycle								
Age	-0.2E2	-2.0	-0.2E2	-1.9	0.3E2	2.4	0.4E2	2.5
Origin [2]								
Business spin out	0.4	1.7	0.2	1.0	-0.4	-2.0	-0.3	-1.8
MBO	0.5	1.8	0.3	1.1	-0.1	-0.5	-0.1	-0.6
New start-up	0.4	1.9	0.2	1.2	-0.3	-1.7	-0.3	-1.9
Unemployment	-0.4E1	-0.5	-0.5E1	-0.5	0.2E1	0.3	0.2E1	0.3
Growth								
Employment 87/90	0.2E3	1.1	0.1E3	0.7	0.1E3	0.6	0.2E3	1.3
Profitability								
Net profits/turnover	0.1E1	3.8	0.1E1	3.5	-0.1E1	-4.7	-0.9E2	-3.4
Trade								
Exports/Turnover	0.1	0.9	0.2	1.4	-0.3E1	-0.2	0.1E1	0.1
Innovation								
Innov. in last 5 yrs	0.1E2	0.01	0.8E2	0.8E2	-0.1	-1.3	-0.2E1	-0.3
Size								
Employment 1990	0.7E3	1.8	0.6E3	1.5	-0.1E2	-3.1	-0.1E2	-3.2
Region								
North	0.8E1	0.9	0.3E1	0.4	-0.4E1	0.5	0.5E2	0.1

Source: SBRC data (1992)

Notes: 1 Omitted variable conventional services; 2 Omitted variable merger or purchase of existing firms $E1 = \times 10{-1}$, $E2 = \times 10{-2}$, etc.

acting as a highly important constraint (rank 8 and 9) relative to firms in the conventional service sector).[3] Conventional manufacturing firms also reveal a tendency to be constrained by finance, although in both cases the coefficients are statistically significant only at the 10 per cent level of significance. The younger the firm, the greater the likelihood of facing severe financial constraints (positive and significant coefficient on ranks 8 and 9).[4] Origin is not significant for rank 9, although coefficients are all negative. However for ranks 8 and 9 together, origin emerges as a negative and significant variable for business spin-out and positive and significant for ranks 0 and 1. Growth of employment is not significant but profitability is highly significant and positive for rank 0 and ranks 0 and 1 together and negative for rank 9, and rank 8 and 9 together, indicating perhaps not surprisingly that financial constraints on expansion are less likely the more profitable the small firm. Trade is not statistically significant and neither is there evidence of a North/South effect. Finally, the larger the firm, the lower the probability of facing a highly important constraint in obtaining finance for expansion.

Table 4.14 shows the regression results for the availability and cost of overdraft finance. Conventional manufacturing firms again display a propensity to face constraints. Age is significant for ranks 0 and 0 and 1 with a negative coefficient indicating that younger companies have an above average probability of overdraft finance being important. Completely new start ups are less likely to face overdraft constraints which is highly important. Of the other independent variables profitability and size are both statistically significant with the expected sign on the coefficients.

FINANCIAL CONSTRAINTS AND THE INTRODUCTION OF NEW TECHNOLOGY

The relative importance of financial constraints

Given the importance of new technologies both in raising efficiency and in improving market competitiveness, it is critically important to assess whether a lack of finance constrains the introduction of new technologies. Moreover because the focus of this chapter is on the high-technology sector, there is an interesting question as to whether high-technology companies face more severe problems than 'conventional' firms. A priori there are grounds for believing this might be the case. Thus, small high-technology firms are unlikely to have a large portfolio of new projects and therefore be limited in the extent

Table 4.14 Regressions results: dependent variable – availability and cost of overdraft finance

Independent		Ranking of importance							
		Unimportant				Highly important			
		0 and 1		0		8 and 9		9	
Variable		Coeff.	t-ratio	Coeff.	t-ratio	Coeff.	t-ratio	Coeff.	t-ratio
Activity[1]	High-tech man.	0.2E1	0.1	0.1	0.6	0.2	1.2	0.2	1.3
	High-tech serv	−0.1E1	−0.1	−0.4E1	0.3	0.1	1.5	0.2	1.5
	Conv. man	−0.1	−1.4	−0.2	−2.2	0.2	2.3	0.2	2.5
Life cycle	Age	−0.3E2	−2.7	−0.3E2	−2.3	0.1E2	0.9	0.7E2	0.5
Origin[2]	Business spin out	0.3	1.7	0.2	1.2	−0.3	1.8	−0.4	−2.1
	MBO	0.4	1.4	0.3	1.3	−0.8E1	−0.3	−0.2	−1.0
	New start up	0.3	1.5	0.2	1.1	−0.2	−1.3	−0.3	−2.1
	Unemployment	0.6E1	0.1	0.2E1	0.3	0.7E1	0.8	0.8E1	0.9
Growth	Employment 87/90	−0.2E3	−0.9	−0.2E3	−1.0	−0.1E3	−0.6	0.4E4	0.2
Profitability	Net profits/turnover	0.1E1	5.3	0.1E1	4.6	−0.9E2	−3.5	−0.1E1	−4.3
Trade	Exports/Turnover	0.1	0.8	0.1	0.6	0.6E1	0.4	−0.9E2	−0.1
Innovation	Innov. in last 5 yrs	−0.1	−1.2	−0.1	−1.4	−0.9E2	−0.1	−0.6EI	−0.7
Size	Employment 1990	0.8E3	2.2	0.6E3	1.7	−0.8E3	−2.0	−0.1E2	−3.0
Region	North	−0.7E1	−0.8	−0.9E1	−0.9	0.4E1	0.5	0.8E1	0.9

Source: SBRC data (1992)
Notes: 1 Omitted variable conventional services; 2 Omitted variable merger or purchase of existing firms E1 = × 10–1, E2 = × 10–2 etc.

Table 4.15 The relative importance of different factors limiting the introduction of new technology (mean rank: 0 = unimportant, 9 = highly important)

Limiting factor	'Conventional' 'Manufacturing'	Services	High-technology 'Manufacturing'	Services
Cost of purchase and installation	6.2	5.6	5.7	5.3
Cost of operational support/services	4.4	4.5	4.3	4.0
Lack of finance	5.1	4.4	4.6	5.0
Difficulty in recruiting skilled staff	3.1	2.7	3.4	2.6
Inability to recruit suitable management	2.6	1.9	2.5	2.2
Problems in training staff	2.3	2.3	1.9	2.0
Resistance of staff	2.0	1.9	2.1	1.4
Unsuitability of premises	1.9	1.4	1.5	1.2

to which it can diversify away some risk by comparison with the larger firm. The new technology may be very specific to the type of product/ service produced by the firm and as such be of little value should the project (or firm) fail. If the new technology is embodied in capital equipment for R&D the ultimate payoff may be very uncertain and unlikely to appeal to risk averse investors. Table 4.15 shows that the cost of purchase and installation is the most important factor limiting the introduction of new technology, particularly for 'conventional' manufacturing (6.2). Lack of finance is ranked second in importance with the 'conventional' manufacturing sector giving a slightly higher ranking than high-technology manufacturing, with the reverse for the services sector.

Regression analysis

A regression analysis was carried out with lack of finance as the dependent variable and a set of independent variables covering the same company characteristics as before. The results are shown in Table 4.16. Lack of finance would appear to be highly important (rank 9) or (8 and 9) for both high-technology and 'conventional' manufacturing (relative to conventional services) in introducing new technology. It is also clear that the younger the firm the more likely finance will be a highly important problem. Business spin-outs and completely new start ups would appear to have a relatively low probability that lack of finance would present a problem for new technology. The higher the profit margin the lower the probability

Table 4.16 Regression results: dependent variable – lack of finance for the introduction of new technology

Independent		Ranking of importance							
		0 and 1		0		8 and 9		9	
Variable		Coeff.	t-ratio	Coeff.	t-ratio	Coeff.	t-ratio	Coeff.	t-ratio
Activity[1]	High-tech man.	−0.3E1	−0.2	−0.4E1	−0.2	0.4	2.2	0.4	2.0
	High-tech serv	0.9E1	0.7	0.1	1.1	0.7E1	0.5	0.1	0.8
	Conv. man	−0.1	−2.0	−0.1	−1.4	0.4	4.2	0.4	4.0
Life cycle	Age	−0.2E2	−1.5	−0.2E2	−1.7	0.4E2	2.8	0.3E2	1.7
Origin[2]	Business spin out	0.6E1	0.3	−0.5E1	−0.2	−0.1	−0.8	−0.3	−1.6
	MBO	0.6	0.2	0.8E2	0.0	−0.1	−0.7	−0.2	−0.9
	New start up	0.8E2	0.0	−0.1	−0.7	−0.1	−1.7	−0.4	−2.0
	Unemployment	0.6E1	0.6	0.1	0.3	0.2E1	0.3	0.1E1	1.6
Growth	Employment 87/90	−0.4E3	−1.6	−0.5E3	−1.7	−0.1E3	−0.6	0.1E4	0.1
Profitability	Net profits/turnover	0.1E1	3.4	0.9E2	2.9	−0.1E1	−3.6	−0.1E1	−4.1
Trade	Exports/Turnover	−0.1	−0.6	−0.8E1	−0.4	−0.5E1	−0.3	−0.6E1	−0.3
Innovation	Innov. in last 5 yrs	−0.5E1	−0.1	−0.2E2	0.0	0.6E1	0.6	−0.3E1	−0.3
Size	Employment 1990	0.1E2	2.9	0.8E3	2.0	−0.2E2	−4.3	−0.3E2	−4.6
Region	North	−0.5E1	−0.5	−0.1	−1.1	0.8E2	0.1	0.7E1	0.7

Source: SBRC data (1992)

Notes: 1 Omitted variable conventional services; 2 Omitted variable merger or purchase of existing firms E1 = × 10^{-1}, E2 = × 10_{-2} etc that finance will be a highly important constraint and lack of finance presents less of a problem as the firm increases in size.

EXTERNAL SOURCES OF FINANCE FOR EXPANSION

This section turns to the question of the extent to which high-technology and 'conventional' firms had sought external finance during the previous three years and the sources used. Table 4.17 shows that 65 per cent of the sample of firms had sought external finance and as suggested in *The State of British Enterprise* (SBRC 1992) this is a somewhat greater proportion than that identified in the Bolton Committee Report. There is little difference between the two sectors although 'conventional' manufacturing and high-technology services show a slightly greater tendency to seek external finance by comparison with firms in the other two sectors.

Table 4.17 Firms seeking to obtain external finance in the last 3 years (%)

	High-technology	'Conventional'
Manufacturing	60	66
Services	68	63

The relative importance of different sources of finance for the high-technology and conventional sectors shows that banks are by far the most significant providers (see Table 4.18). High-technology manufacturing firms obtain a mean percentage share of 65 per cent compared with 61 per cent for 'conventional' manufacturing and only 51 per cent for high-technology services. Hire purchase or leasing provided 19 per cent of finance for 'conventional' manufacturing and less than half this (8 per cent) for high-technology manufacturing. There was little difference between high-technology and 'conventional' small firms in the proportion of finance raised from working shareholders and partners. However, there was some evidence of rather more 'angel' finance for high-technology companies and indeed the proportion of external finance from private individuals was very similar to that provided by venture capital firms.

Further analysis of the sources of funding for small high-technology firms through from their early stage start-up period to maturity not only indicates a clear life cycle phenomenon but also important differences between high-technology firms and 'conventional' firms (see Table 4.19).

Table 4.18 Sources of additional finance: mean % share by source of finance

Source of finance	'Conventional' Manufacturing	Services	High-technology Manufacturing	Services
Banks	60.9	57.6	64.7	51.3
Venture capital	3.0	2.4	2.5	3.9
Hire purchase/leasing	18.6	12.6	8.3	14.9
Factoring	3.4	4.0	1.9	2.7
Customers/suppliers	1.9	2.3	1.5	2.7
Partners/working shareholders	4.5	10.6	5.5	11.8
Other private individuals	1.3	1.4	2.8	3.9
Other sources	4.1	6.0	13.0	6.0

In the initial period following start up, a significantly higher proportion of finance used by high-technology companies comes from banks. The proportion of finance from Venture Capital firms and customers/suppliers is also noticeably higher for high-technology firms. Over time the proportion of finance from banks gradually increases for both high-technology firms and 'conventional' and for mature firms there is little difference between the two sectors. Hire purchase also increases as the small firms mature but tends to be a more important source in the 'conventional' sector. Similarly financial support from partners/shareholders gradually diminishes in importance for both sectors, although there is some volatility as sample sizes fall with age.

CONCLUSIONS

This chapter has presented new evidence on the importance of financial constraints facing small firms with particular focus on the high-technology sectors. Importantly an attempt has been made to test whether small high-technology firms differ from small firms in the 'conventional' sector in the nature and severity of the financial constraints they confront.

At the start up phase, about one-sixth of small high-technology firms in the face-to-face interview survey faced major problems obtaining and managing finance. Significant as this is however, major problems were equally likely to be confronted with respect to product development and marketing.

Within the range of issues raised under finance, management of cash flow and raising equity and debt finance were cited most frequently as presenting major problems. The main source of finance

Table 4.19 Changing sources of 'additional' finance through the life cycle of high-technology and 'conventional' companies: mean % share by source of finance

Company age	Banks	Venture capital	Hire purchase	Factoring	Customers suppliers	Partners share	Indiv.	Other
Conventional								
< 5	45.6	4.0	11.9	10.4	1.1	10.8	4.6	6.2
6–8	59.1	2.4	14.6	3.7	1.2	9.3	0.7	4.7
9–14	60.4	2.1	16.9	3.5	3.6	7.5	1.0	3.7
15–24	60.8	2.7	16.8	2.4	2.4	4.1	0.8	6.1
25–49	65.1	1.3	20.1	2.6	0.8	4.9	0.2	5.0
50+	62.3	4.1	16.3	1.2	2.5	6.0	0.6	4.3
High-technology								
< 5	53.1	7.0	6.2	5.6	4.3	8.0	3.2	10.7
6–8	45.3	4.1	16.2	2.6	1.3	14.4	6.4	10.6
9–14	56.9	0.9	17.3	1.0	1.1	5.5	2.4	7.1
15–24	72.9	6.7	8.2	0.5	2.1	4.7	–	11.6
25–49	66.6	2.3	3.4	7.7	–	12.3	–	–
50+	61.7	–	23.0	0.0	10.3	–	–	5.0

Source: SBRC survey (1992)

at start up was personal finance, with 44 per cent of companies claiming that this was of overwhelming importance in their start up capital. However one-third of companies relied mainly on external finance, particularly spin-offs from universities/HEIs where over half relied mainly on external capital. Bank finance was relatively unimportant (7 per cent) and Venture Capital firms provided some 10 per cent of start up capital.

Analysis of subsequent company growth, as measured by employment, change provided support for the view that problems with initial capitalization were associated with below average growth.

In the period of early growth and developing maturity, the availability and cost of finance for expansion presents the most important constraint facing high-technology firms but it is only marginally more severe than that facing 'conventional' small firms. The crucially important *distinguishing* constraint facing the high-technology firm is the dearth of marketing and sales skills. In addition market conditions faced by 'conventional' firms were shown to be much more severe as a constraint in securing business objectives than those confronting the average high-technology firm. In general, a striking feature of the analysis is the similarity in terms of frequency and severity of problems facing high-technology and conventional firms.

There is some evidence that faster growing high-technology manufacturing companies (in terms of sales) are more likely to face severe financial constraints than more slowly growing companies. For small high-technology firms in the service sector the evidence is less conclusive and financial constraints are more likely to be regarded as highly important when firm growth is measured by sales, but not when measured by employment.

The regression analysis indicates that small high-technology firms do have a relatively high probability of facing financial constraints by comparison with the 'conventional' services sector but not by comparison with the 'conventional' manufacturing sector.

The regression analysis also suggests that serious finance constraints for expansion are more likely to be faced by younger firms and firms in the conventional 'manufacturing sector'. More profitable firms and larger firms are less likely to face major problems. Whether a firm innovates or not seems unimportant; as a factor influencing firms problems in seeking finance, faster growth of employment is not a significant factor. There is some support from the regression analysis that the origin of the company is important, with business spin-offs and completely new start ups less likely to face highly important

140 *Barry Moore*

finance problems than small firms arising from a merger of existing firms or started as a consequence of unemployment of the founder.

The regression results for the availability and cost of overdraft finances again suggest problems not only for conventional manufacturing firms, but also for high-technology manufacturing firms. Increased profitability and size both reduce the probability of serious problems with overdraft finance.

Lack of finance ranks highly as a factor limiting the introduction of new technology by comparison with a range of other factors such as suitable staff, premises, and so on, but there is little by way of differentiation between high-technology and conventional small firms. The regression analysis reveals severe problems for conventional manufacturing and for high-technology manufacturing. The more mature firm also faces more severe problems with finance by comparison with younger firms. New start ups and business spin-outs are less likely to face problems in obtaining finance by comparison with start ups from other origins. Larger firms and more profitable firms are also less likely to experience lack of finance as a highly important problem by comparison with smaller less profitable firms.

Finally, both high-technology and conventional small firms show a strong tendency to seek external finance following start up. Not surprisingly banks are the most important source for all small firms although small high-technology service companies rely on them somewhat less than small firms in the other sectors. Venture capital is used by a minority of firms and there is little differentiation between the high-technology and 'conventional' sectors.

APPENDIX

Basic characteristics of the SBRC high-technology data base

The data base derived from the postal survey

The SBRC high-technology data base includes within it 292 small firms classified as high-technology small firms. Table 4.A1 shows the breakdown of these firms by activity. Over half the companies are in computer services within the business services sector. The other two significantly represented sectors are in manufacturing and are electrical and electronic engineering and instrument engineering.

Table 4.A1 High-technology firms by main activity

Activity	Number	%
Chemical industry	12	4
Manufacture of office and data processing machinery	7	2
Electrical and electronic engineering	39	13
Manufacture of other transport equipment	2	1
Instrument engineering	36	12
Postal services and telecommunications	5	2
Business services	172	59
Research and development	19	7
Total	292	100

The age distribution of businesses in the study sample is summarized in Table 4.A2. Very few firms (less than 3 per cent) are less than two years of age but nearly one-fifth of the sample are aged under five. Over 25 per cent are relatively mature being aged over fifteen. By comparison with the 'conventional' small firms the main difference is that there are a relatively smaller proportion at the more mature end of the age spectrum and a relatively higher proportion of younger firms.

Table 4.A2 Age of small high-technology businesses and small firms in the conventional sector

Age of firm	'Conventional' Number	%	High-technology Number	%
0–2	19	1	8	3
3–5	218	13	53	19
6–8	272	16	59	21
9–14	379	22	88	31
15–24	344	20	37	13
25–49	233	14	29	10
50+	223	13	12	4
Total	1688	100	286	100

Table 4.A3 shows the size distribution of the sample of high-technology firms in the 'conventional' sector.

In terms of employment, a higher proportion of the smallest companies (0–9) are in the high-technology and 'conventional' services sectors, 41 per cent and 36 per cent respectively, by comparison with about 20 per cent in each of the manufacturing sectors. By contrast, the larger companies with 100 or more employees tend to be concentrated more in the manufacturing sectors. Few firms have a turnover exceeding £10m in the high-technology sectors (21) and the great majority are divided broadly equally in the two bands £100k–£1m and £1–10m.

Table 4.A3 Size distribution of firms in the high-technology and 'conventional' sector by employment and turnover in 1990 (% total sample)

| | 'Conventional' | | | | Hi-technology | | | | | |
| | Manfg. | | Services | | Manfg. | | Services | | Total | |
	No.	%	No.	%	No.	%	No.	%	No.	%
Employment										
0–9	173	19	252	36	18	20	77	41	520	28
10–49	357	40	238	34	30	33	67	36	694	37
50–99	174	19	103	15	20	22	25	13	322	17
100–199	94	10	61	9	7	8	10	15	172	9
200+	100	11	44	6	15	17	9	5	168	9
Total	898	100	698	100	90	100	188	100	1876	100
Turnover £000s										
< 100	26	3	43	6	3	3	15	8	87	5
100–999	348	39	315	46	35	40	92	51	791	43
1000–9999	418	47	283	41	36	41	67	37	805	44
10,000–49,000	87	10	36	5	12	14	7	4	142	8
50,000+	6	1	4	1	2	2			12	1
Total	885	100	681	100	88	100	181	100	1837	100

The database derived from the 89 high-technology firms interviewed

The main activities by age of the sample firms in the interview survey are shown in Table 4.A4.

Table 4.A4 Distribution of high-technology firms by their main activity and age (years)

| Activity | 0–3 | | 3–5 | | 5–10 | | 10+ | | Total | |
	No	%	No	%	No	%	No	%	No	%
Manufacturing										
Computer hardware	3	3.3	1	1.1	7	7.7	1	1.1	12	13.5
Electronics	3	3.3	–	–	2	2.2	6	6.6	11	12.4
Car/Aerospace	3	3.3	–	–	–	–	1	1.1	4	4.5
Medical	1	1.1	3	3.3	4	4.4	6	6.1	14	15.7
Other	–	–	–	–	5	5.5	4	4.4	9	10.1
Services										
Computer software	3	3.3	4	4.4	13	14.6	8	8.9	28	31.5
Other business	1	1.1	2	2.2	5	5.5	4	4.4	11	12.4
Total	14	15.7	9	10.1	36	40.4	30	33.7	89	100

Over half the firms are in the manufacturing sector (56.1 per cent) and almost three-quarters (74.1 per cent) are aged over five years.

Table 4.A5 shows that nearly 16 per cent of the sample were very small firms employing less than six persons with a further 28 per cent (twenty-five companies) employing between six and ten employees.

Table 4.A5 Size of business in 1991 (number of employees)

Size	Number	%
1–5	14	16
6–10	25	28
11–20	16	18
20–50	29	33
50+	5	6
Total	89	100

In the follow up telephone enquiry 43 of the 89 firms responded to a short questionnaire focused on seed capital finance.

NOTES

1 The first database is derived from 2,028 usable responses to a national postal enquiry of just over 6,000 small companies and included 292 high-technology firms according to the definition of Butchart (1987). The second database derives from a face-to-face interview programme of eighty-nine small high-technology firms and a short follow-up telephone survey of issues relating to seed capital finance. Key characteristics of these two databases are provided in the Appendix to this paper.
2 This section draws mainly on the results of the eighty-nine face-to-face interviews with small high-technology firms. Some of the results of which were presented to Anglo-German Foundation sponsored Workshop in Oxford in 1992 based on a short paper, B. Moore, R. Moore and N. Sedaghat 'Early stage finance for small high-technology companies: a preliminary note'.
3 Experiments omitting other sectors (e.g. conventional manufacturing) did not reveal high-technology manufacturing or services to be statistically significant sectors.
4 Age is measured by year of start, implying younger firms as year of start increases.

REFERENCES

Advisory Council on Science and Technology (ACOST) (1990) *The Enterprise Challenge: Overcoming Barriers to Growth in Small Firms*, HMSO, London.
Binks, M.R., Ennew, C.T. and Reed, G.V. (1992) 'Information asymmetries and the provision of finance to small firms', *International Small Business Journal*, 11, 1.

Bullock, M. (1983) *Academic Enterprise, Industrial Innovation and the Development of High Technology Financing in the United States*, Brand Brothers.

Butchart, T. (1987) 'A new UK definition of the high technology industries', *Economic Trends*, HMSO, No 400 London.

Department of Trade and Industry (1991) *Constraints on the Growth of Small Firms*, HMSO, London.

Van Glinow, M.A. and Mohrman, S.A. (1990) *Managing Complexity in High Technology Organisations*, Oxford University Press, Oxford.

Hall, G. (1992) 'Reasons for insolvency amongst small firms. A review and fresh evidence', *Small Business Economics*, 4, 237–44.

Hall, G. and Young, B. (1991) 'Factors associated with insolvency amongst small firms', *International Small Business Journal*, 9, 54–64.

Hunsdiek, D. and Albach, H. (1988) 'Financing the start-up and growth of NTBFs in Germany', in *New Technology Based Firms in Britain and Germany*, Anglo-German Foundation.

Hutchinson, R.W. and McKillop, D.G. (1992) 'Banks and small to medium sized business financing in the United Kingdom: Some general issues', *National Westminster Bank Quarterly Review*.

Kay, J. (1992) 'Innovation in corporate strategy' in Bowen, A. and Ricketts, M. *Stimulating Innovation in Industry*, NEDO.

Moore, B., Moore, R. and Sedaghat, N. (1992) 'Early stage finance for small high-technology companies: a preliminary note', paper presented to a DTI/Anglo-German Foundation workshop in Oxford.

Murray, G. (1992) 'Change and maturity in the UK Venture capital industry, 1991–95', Warwick Business School, Coventry.

Oakey, R. (1984) *High Technology Small Firms*, Frances Pinter, London.

Oakey, R.O., Faulkner, W., Cooper, S.Y. and Walsh, V. (1990) *New Firms in the Biotechnology Industry: Their Contribution to Innovation and Growth*, Pinter, London.

Roberts, E.B. (1991) *Entrepreneurs in High-technology*, Oxford University Press, Oxford.

Roberts, E.B. and Hauptman, O. (1987) 'The financing threshold effect on success and failure of biomedical and pharmaceutical start-ups', *Management Science*, 33(3).

Slatter, S. (1992) *Gambling on Growth, How to Manage the Small High-tech Firm*, John Wiley.

Small Business Research Centre (1992) *The State of British Enterprise: Growth, Innovation and Competitive Advantage in Small and Medium-sized Firms*, SBRC, University of Cambridge.

Storey, D. and Strange, A. (1992) *Entrepreneurship in Cleveland 1979–1989*, Centre for Small and Medium sized enterprises, Warwick Business School, University of Warwick, Employment Department Research Series No. 3.

Storey, D. Watson, R. and Wynarczyk, P. (1989) 'Fast growth of small business', Research Paper Number 67, Department of Employment.

Stoy-Hayward (1992) 'Report of Survey' in Scotlands Business birth rate, Scottish Enterprise.

Tiler, C., Metcalfe, S. and Connell, D. (1990) 'The management of growth: negotiating transitions', paper presented at a symposium 'Growth and development of small high tech businesses', Cranfield Institute of Technology, April.

5 Raising capital for the ethnic minority small firm

Trevor Jones, David McEvoy and Giles Barrett

UNEQUAL ENTERPRISE

Racial disadvantage is now widely acknowledged as part of the fabric of British life. During the 1980s, when many members of ethnic minority communities sought self-advancement by joining the swelling ranks of the self-employed, it became evident that this area was no more a level playing field than any other. Among the many barriers to be hurdled was the problem of raising capital, a perennial problem for new and small businesses as a whole, but seemingly more than normally intractable for entrepreneurs of immigrant origin, especially those such as West Indians, Africans, Indians, Pakistanis and Bangladeshis who are racially labelled.

The present chapter examines this problem with reference to interviews carried out in 1990–1 with 403 small firms, 178 Asian-owned, 54 Afro-Caribbean and 171 white in 15 locations throughout England.[1] Respondents were asked about their sources of start up and subsequent capital and the approximate proportion contributed by each. No question was asked about precise sums of money involved, since previous experience has taught that such questions may be sensitive, especially when, as in this case, interviews have not been pre-arranged and confidence has to be built up from a cold start. On occasions such information was volunteered without prompting, but it was not offered sufficiently to enable us to make direct comparisons with such researchers as Wilson and Stanworth (1987), one of whose important findings was that Asians have access to greater sums of capital than Afro-Caribbeans. Nevertheless, we are in a position to comment on overall patterns of sourcing on the basis of a sample rather larger than most used hitherto in this field. In practice we obtained definite responses on this item from 316 of the interviews, the missing 87 usually being interviews where the owner was absent and the manager/representative unable or unwilling to divulge information.

In addition to funding methods, we also obtained information on the personal characteristics, career histories, family background and business motivations of the entrepreneurs, together with a range of details of their premises, clientele, suppliers, working practices, problems and performance. This provides contextual support for the data on funding, enabling us to throw some light on what kinds of firms and owners obtain what kinds of finance; and the effects of different modes of funding for firms' performance. Before reporting these findings, however, we need to give a brief account of ethnic minority business in Britain and of the special funding problems facing immigrant-origin firms.

ETHNIC MINORITIES IN BUSINESS

Among the many novelties of the entrepreneurial 1980s, perhaps the most immediately surprising was the emergence and rise of the ethnic minority-owned firm. In an age where official emphasis was newly placed on the new and small firm as a central means of economic regeneration, it soon became apparent that the most conspicuously 'entrepreneurial' sections of the population were members of immigrant-origin groups. Despite often entering the country as recruits for low-level jobs increasingly shunned by the indigenous working class, and notwithstanding racial discrimination, cultural alienation, language barriers and the trauma of displacement, many such groups have forged ahead as entrepreneurs. According to Curran and Burrows (1988), Britain in the early 1980s was home to several ethnic minority groups for whom entrepreneurship is a distinct occupational specialization, with rates of self-employment and business ownership significantly ahead of those of the population at large. Among those numbered in Table 5.1 are Cypriots, Maltese and Gibraltarians and also Chinese, a component (possibly the major one) of the 'Rest of the world' category.

Most outstanding of all are those described as Indian and Pakistani–Bangladeshi, now in common parlance bracketed together under the shorthand term 'Asian', an inexact but convenient codeword describing all those of Indian sub-continental origin, including those who arrived here via some intervening location such as East Africa. Though in proportionate terms Asian self-employment rates fall short of the afore-mentioned groups, they are well above those of the white native-origin population, a gap which has widened still

Table 5.1 Self-employment and business ownership rates

Ethnic origin	Small business owners %	Self-employed %
Native white	2.4	6.0
Other European	3.6	8.4
Cypriot, Maltese and Gibraltarians	16.0	13.4
Indian sub-continent	5.7	7.9
Afro-Caribbean	0.2	3.1
Rest of world	7.0	6.3

Source: after Curran and Burrows 1988, data for 1981–4

further since Curran and Burrows' findings (Ward 1991). Moreover the Asian impact on the entrepreneurial map of Britain is far greater in aggregate terms than that of any other ethnic minority group. Whereas the Chinese, for example, are shown by the 1991 Census to make up only 0.3 per cent of the resident population of England and Wales, people of Indian–Pakistani–Bangladeshi origin are almost ten times as numerous. This group of Asian entrepreneurs is thus likely to be very numerous in its own right as well as proportionately over-represented, a particularly remarkable occurence in that less than three decades ago there were only a comparative handful of Asian-owned firms in the country. Clearly we are witnessing the flowering of what Light and Bonacich (1988) call an *entrepreneurial minority*, a group which has succeeded in inverting all the normal laws of minority disadvantage by cultivating entrepreneurship as an occupational specialization. Far from an underclass or a sub-proletariat, Asians in Britain are ostensibly taking on a distinctly petty bourgeois character.Indeed to judge from the publicity enjoyed by Britain's Asian millionaires (*Daily Express* 27 June 1990, *Today* 25 August 1990, *Asian Business* 13–26 September 1991, *Mail on Sunday* 5 July 1992), this form of social mobility need not be confined merely to entry into the *petty* bourgeoisie.

Very frequently in the ethnic minority enterprise literature, any consideration of Asian entrepreneurial self-advancement leads to comparisons with Afro-Caribbeans, the other major ethnic minority population created by the great migrations from the 'New Commonwealth' in the 1950s and 1960s. Such comparisons are invariably to the detriment of the latter. As confirmed by Table 1, Afro-Caribbean rates of self-employment/business ownership lag behind both national rates and, even more emphatically, Asian rates. Although Afro-Caribbean business formation is undoubtedly procceding (partly in response to Scarman's [1981] exhortations and to enterprise policy initiatives), they are comparatively late arrivals on the entrepreneurial scene and are usually depicted as beset by particularly impenetrable barriers to business entry and viability (Ward and Reeves 1980, Wilson 1983, Reeves and Ward 1984, Wilson and Stanworth 1987, Jones *et al.* 1989). In crude quantitative terms there is no doubt of the highly distinctive entrepreneurial profiles exhibited by these two ethnic minorities, distinct not only from one another but from the general population also. This inter-ethnic differential is reminiscent of that observed in the USA between recognized entrepreneurial minorities such as Chinese, Japanese and Koreans on the one hand and African Americans on the other, the latter seemingly consigned to a permanent ghetto existence, with entrepreneurship offering a viable escape route for only a tiny minority (Bonney 1975, Cashmore 1991).

Combing through the British literature on this Afro-Asian gap, we find that a recurring theme is *differential access to business resources*, with Asians far more successful than Afro-Caribbeans in mobilizing key resources such as capital, labour and markets (Soar 1991). This clearly needs to be put into historical perspective. At the outset we might suppose that both Asians and Afro-Caribbeans would be severely disadvantaged in respect of business resources, given the racist nature of British society and the circumstances of their entry into it. Initially recruited into and typecast for certain low level occupations and a restricted range of industries, they were unusually concentrated in the lower strata of the working class, their occupational mobility impeded by all manner of discriminatory and exclusionary practices (Smith 1976, Phizacklea and Miles 1980, Brown 1984, Ohri and Faruqi 1988). With bitter irony, the very jobs on which they had become reliant proved to be the most expendable in the subsequent wave of deindustrialization and public sector cuts. Ethnic minority vulnerability to job loss was even greater than that among the general population which, in the usual absence of alternative job opportunities, created mounting pressure for entry into self-employment as a survival strategy (Patel 1991). This is confirmed in the present survey, where almost one-third of Asian, Afro-Caribbean and white respondents gave unemployment or some other form of labour market 'push' as their principal motive for business entry (see also Jones *et al.* 1989, Ram and Sparrow 1992).

As Ward (1991) argues, the response of the two ethnic minorities to these straitened circumstances have differed sharply. Whereas a great many Asians have successfully cultivated self-employment as a viable alternative, Afro-Caribbeans seem less willing or able to follow this strategy. In Ward's words:

> the impetus to business ownership among the South Asian popula-
> tion was the sharp and sustained rise in unemployment of the late
> 1970s and 1980s, rather than a predisposition for entrepreneurship
> . . . [but] . . . while Afro-Caribbeans shared the experience of
> Asians of large-scale redundancy, they have not been able to
> respond by a significant move into self-employment.
>
> (Ward 1991: 58)

On the one hand, racial disadvantage appears to have provided a positive stimulus for entrepreneurial self-advancement (see Weber [1976] for the classic exposition of the links between outsider status and entrepreneurship); on the other hand, it appears as part of a self-perpetuating cycle of unemployment and poverty. In attempting to explain these differing adaptations, the ethnic minority enterprise discourse tends to lean rather heavily on culturalist interpretations. From the earliest contributions by Patterson (1969) and Hiro (1971), many commentators quite correctly recognize the distinctive histo-rical experience of the two groups and the way these have left their legacy in the social attitudes and behaviour of their present day communities in Britain. This is applicable right across the board, but it rings with special resonance when applied to entreprencurial behaviour. In the Asian case, tight-knit communal and family soli-darity provides a highly effective insider network for the mobilization of capital, labour, markets, skills, experience, inter-firm linkages and other competitive assets. Ram (1992) writes of Asian family networks as a means of 'negotiating racism', a source of collective strength acting as a defence against a hostile external world or even as a means of counter-attack. Additional business dynamism is conferred by various traditional values, secular and religious, which are held to be congenitally 'Asian' properties and to be conducive to a pro-entrepreneurial world view. Thus Werbner (1984) writes of a 'Pakistani ethos' of industriousness, frugality, self-help and deferred gratifica-tion; Helweg (1986) writes of the essentially hard-working predisposi-tion of the Sikh community (see also Ballard and Ballard 1977 and Bains 1988 for a broadly similar interpretation of Sikhs in Britain); Patel (1991) of an entrepreneurial 'predeliction' allegedly peculiar to Asians. In the Afro-Caribbean case the assumed absence of these values and institutions (or, for West Indians, their extinction under slavery) has led to a relative absence, both of entrepreneurial resources and of motivation. Table 5.2 summarizes in highly simplified

Table 5.2 Contrasting images of Afro-Caribbeans and Asians in business

	Asian	Afro-Caribbean
Ethnic resources		
Personal motivation	Strong	Weak
Family support	Strong	Weak
Communal support	Strong	Weak
Class resources		
Business family background	Uneven but growing	Weak
Academic qualifications	Moderate, growing	Weak
Artisanal skills	Weak	Moderate to good
Opportunity structure		
Racist labour market (negative)	Strong	Strong
Discrimination in business (negative)	Strong, waning	Strong, persisting
Niche markets	Strong	Weak

form the standard assumptions about the manner in which ethnic traits are likely to influence the mobilization of resources relevant to business ownership.

CONTRASTING METHODS OF BUSINESS FUNDING

When we concentrate on the means by which ethnic minority firms raise capital, we find that Ward's (1991) view of events seems completely apt, with Asians ostensibly much more effective than Afro-Caribbeans in generating funding. At first sight this is surprising. In the absence of any compensating conditions it might be supposed that capital formation is an area in which *both* ethnic minorities would be more than usually handicapped, both by class position and by racial discrimination itself. With regard to class, we have already noted the initial concentration of the two groups in the lower strata of the working class, a position unlikely to equip potential recruits with the resources they need for the demanding career of business owner. Capital is the most conspicuous of the resources they will lack. A career history of manual work and low earnings is hardly a sound basis on which to accumulate cash savings or to acquire property as collateral for bank credit. This deficiency will be amplified by other reasons, less tangible but no less decisive, such as lack of information about sources of institutional finance, lack of self-confidence in approaching banks and lack of knowledge about how to present a viable business plan.

In a very real sense, these disadvantages are common to all those from a working-class background, irrespective of ethnicity. The small business literature consistently emphasizes the material and other advantages to entrepreneurs of a middle-class background, membership of a business family itself being seen as the most potent resource base of all (Bechofer *et al.* 1974, Curran 1986). Factors like these are not ethnic-specific and elsewhere we have warned against a tendency to *particularize* ethnic minority business owners as in every respect special cases (Jones *et al.* 1992a, 1992b). Such a perspective abstracts them from their true context, plays down the effects of conditions common to all small owners and, if over-emphasized, depicts them virtually as exotic abnormalities. Having insisted on this, however, we would also stress that Afro-Caribbeans and Asians experience an aggravated and racialized version of this common disadvantage. We see this as operating on two fronts. First, being inordinately over-exposed to job loss and hence pressured into self-employment, they are automatically over-exposed to the problems inherent in business entry from the working class. Second, because of the persistence of discrimination and of negative racist stereotypes, their occupational progression has been consistently impeded. This extends to their attempts to enter self-employment itself, where discriminatory barriers erected by customers, suppliers, local authorities and financial institutions are now extensively documented (Ward and Reeves 1980, Wilson 1983, Creed and Ward 1987, Wilson and Stanworth 1987, Patel 1988, Jones *et al.* 1989, 1992b). Moreover, in relation to banks and other lending agencies, we should note ethnic minority entrepreneurs' own inhibitions, their frequent alienation from what they see as a 'white' institution, run by whites for whites and from which they themselves are excluded, psychologically if not in practice. For Asians, this non-communication is compounded by language barriers. Only one in ten of our present Asian sample were born in Britain, implying a heavy preponderance of Asian business owners for whom English is not their first language, a preponderance destined to disappear in the future, but strongly operative in the present.

The recent Bank of Credit and Commerce International fiasco, which seriously damaged scores of Asian businesses, served also to expose the weaknesses of the conventional high street banks *vis-à-vis* ethnic minority customers. According to Hasmukh Shah, spokesperson for the World Council of Hindus, the attraction of BCCI was that it was presented as an Asian bank and managed by Asians for

Asians. It offered dialogue in the client's mother tongue and was sympathetic to the needs of the small Asian depositor. Naturally 'people in our community turned to it and used it as a bank in the normal way' (Shah 1991). In effect BCCI flourished, albeit ephemerally, in a yawning vacuum created by the deficiencies of the mainstream banking system.

As drama and tragedy, the BCCI crash was highly newsworthy and served to bring the sufferings of small Asian customers vividly before the public gaze. Even so, a true sense of perspective would acknowledge this as a 'one-off' event, highly damaging certainly and indicative of the vulnerability of the Asian economic position, but hardly representative of their long-term experience. Beginning with Ward and Reeves (1980), most studies in the field have concluded that, in comparison to Afro-Caribbeans, Asian entrepreneurs have responded surprisingly well to the problem of raising capital, successfully devising strategies for circumventing or eliminating barriers. Asian entrepreneurial entry, viability and expansion are aided, it is argued, by two major advantages:

1 preferential access to informal sources of low-cost funding from within their families and communities;
2 a gradual improvement in the general credit rating of Asian business loan applicants.

This latter view put forward enabled Ward and Reeves (1980) to argue that Asian credit applicants are relatively advantaged in comparison to Afro-Caribbeans. We would add a word of caution here.

Such arguments can all too readily be translated into 'common-sense' stereotypes of Asians as enjoying some kind of unfair and privileged position which makes it impossible for others to compete with them on equal terms. This was manifest in the attitudes of some of our white respondents. Not entirely untypical was the resentful Luton shopkeeper who claimed that 'Banks are too prejudicial against whites. Reputation of Asians in business gives them the advantage in getting a loan'. In all fairness it must be said that there were many white entrepreneurs aware that the true situation is the reverse, but care needs to be taken nevertheless to qualify the argument. The most accurate interpretation would perhaps be that a kind of cumulative process has begun to take effect. Initially, the emergence of Asian business in Britain was largely self-financed through personal savings and pooling of informal resources (Dahya 1974). Having thus achieved a track record and a widespread reputation as business operators, Asian business applicants encountered diminishing resistance from formal sources like the high street banks. In effect, an unfavourable racist stereotype in which ethnic minority members are automatically seen as not credible in the role of entrepreneur is replaced by a new image, still stereotypical, but none the less rather more favourable. Afro-Caribbeans have been largely unable to follow this trajectory, first, because they lack the solidary family-community structures which underpin the process of informal financing; and second, because they continue to encounter powerful resistance from institutional lenders. We shall examine these propositions in the following sections by reference to the responses given by our 410 interviewees.

INFORMAL SOURCES OF FINANCE

As a first stage we examine the relative importance of informal sourcing as opposed to commercial sources of credit. Table 5.3 gives details of those respondents who obtained more than 75 per cent of their start-up funding from either source. As can be seen from the final column, these comprise just under three-quarters of the total sample, the remaining one-quarter being firms who raised their start up from a more even mixture of formal and informal sources.

At first sight this table is somewhat difficult to interpret, giving contradictory signals, some supportive of the conventional wisdom, others less so. It is immediately apparent that each of the three ethnic origin groups exhibits its own distinct pattern, with Afro-Caribbeans emphatically the most dependent on the informal sector and Asians even more definitely least so. This could be taken as approximate confirmation of Afro-Caribbeans' continuing difficulties with banks and of an easing of this problem for Asians. Equally it could be taken to mean that the Asian propensity for self-finance has been exaggerated in the literature; or again that it was strongly operative in the past, but is now no longer a powerful imperative. This might also be inferred from the further finding that our Asian respondents' access to bank sourcing is on a par with that of white respondents.

Table 5.3 Formal and informal start-up sources

More than 75 % from	White	Asian	Afro-Cbn	All
		(per cent of firms)		
Banks and other financial institutions	38.5	39.9	22.7	37.0
Non-market sources	41.5	29.4	50.0	37.0

Moreover, Asians are the only one of the three groups to display a greater orientation towards formal as opposed to informal sourcing. Judging from our own 1978 findings, which showed very low rates of Asian bank borrowing, this heralds a profound shift from informal to formal sector (Aldrich *et al.* 1981, 1984). Here we would concur with Deakins *et al.* (1992) that significant change is occurring, with bank finance now becoming the dominant mode of Asian enterprise funding.

Shifting the focus away from ethnic specificities, we may also conclude from Table 5.3 that the overall reliance on informal sourcing is rather heavy, with over one-third of the total sample more than 75 per cent dependent on this method. White respondents themselves have a high rate of informal sourcing, leading us to suppose that this practice is an established small business tradition, a universal norm rather than a special immigrant adaptation. This supposition is in line with numerous previous findings. Woodcock (1986) notes that self-financing has traditionally provided the main channel for independent business in Britain, a time-honoured custom and practice long pre-dating the influx of immigrant entrepreneurs. Similarly Jennings (1991) verifies that no less than 60 per cent of new-starts are wholly or mainly informally capitalized. Further reinforcement is provided by Mason *et al.* (1991), who also highlight the decisive importance of self-finance at the pre-entry and entry stages.

Interestingly, the strongest case for resisting the temptation to borrow from the formal credit market was made out by a white respondent, a woman over 80 years old who has been operating her shop in Bexley for 55 years: 'What I never had in money I did without. So many businesses fail because they start off in debt.' From this and similar 'never a borrower or a lender be' observations by long-serving white business owners, it is tempting to conclude that, contrary to widely canvassed beliefs, immigrant entrepreneurs have no special monopoly on such values as self-reliance, frugality and abstemiousness. These ideas are equally at home in the traditional European small business culture, the ethos which historically has sustained its participants and given meaning to an often materially unrewarding role (Bechofer and Elliott 1978, Aldrich *et al.* 1986). Indeed a recurring theme throughout the present research is the high frequency with which respondents of all ethnic backgrounds tend to abstain from usage of the formal insititutions of the market and of state bureaucracy (see Table 5.4).

In part this can certainly be explained in cultural and behavioural terms. If, as was the case for at least a third of the present sample,

Table 5.4 Non-usage of formal business practices and agencies

Per cent firms not using:	White	Asian	Afro-Cbn	All
Accountant	5.6	2.4	10.6	4.8
Solicitor	12.5	12.3	15.6	12.8
Advertising	50.6	39.5	52.0	48.6
Formal premises search[1]	61.9	52.3	57.1	56.9
Trade association	69.9	77.8	86.0	75.7
Enterprise support	91.8	87.6	79.6	88.1

Note: 1 Formal search methods for premises are defined here as the use of agencies and published materials such as those used by estate agents, the local council and newspapers to obtain knowledge about business properties. This is in contrast to relatively more casual methods such as knowledge of vacancies obtained from friends.

the individual's driving motive is self-determination and control over the everyday work situation (Jones *et al.* 1992a), then the less the reliance on external agencies the better: or as several respondents put it, 'I don't want the hassle' or words to that effect (cf. Scase and Goffee 1980). In terms of pure capitalist logic, this would presumably be 'irrational' behaviour, but the majority of our respondents evidently do not see themselves as capitalists (Jones *et al.* 1992a). In any case there *are* perfectly sound (if short-term) economic reasons for eschewing the services of market institutions, bankers in particular – they are very costly in comparison to the non-market alternatives. At a time when the financial pages of the British press are constantly agonising over the impact of high interest rates and over accusations of exorbitant charges by the high street banks (Home Office 1991), the attractions of self-finance or interest-free loans from family members have never been stronger.

In addition to all this, we should remind ourselves of the obvious point that reliance on non-market sources is a matter of rejection by institutional lenders as well as spontaneous abstention on the part of entrepreneurs. Various authorities are agreed that the high street banks' preoccupation with security as a cardinal criterion creates a serious shortage of venture capital for would-be business entrants (Woodcock 1986, ACOST 1990, Mason *et al.* 1991). They further observe that barriers to borrowing tend to act in rather subtle ways, more often arising from lack of communication than from bankers' outright recalcitrance. Woodcock (1986) argues that there is a continuing lack of knowledge on the part of small firms about possible funding options and Mason *et al.* (1991) observe that entrepreneurs often have difficulty in convincingly conveying their specialized knowledge to a bank manager. This may well be because, as Scase and Goffee (1980) and Mason *et al.* (1991) argue, the small owner's own interests tend to lie in making, selling and doing rather than in the purely financial side of the business.

Where precisely does the small firm find the necessary alternative non-market sources of start-up capital? In practice the answers to this question are quite varied, with small firms often using a mix of financing methods, but with a heavy dependence on personal savings.

PERSONAL SAVINGS

Confirmation of the above is furnished by Table 5.5, which gives considerable prominence to individual saving as a method of raising start-up finance. It has been frequently argued that the capacity to generate this kind of self-finance is particularly high among immigrant groups, especially those whose original intention was that of temporary migration for the exclusive purpose of economic gain (see Bonacich 1973 on the 'sojourner' mentality and the manner in which this shapes the economic behaviour of migrant groups who see their exile as temporary). Many elements of this logic have been applied to Asians in Britain by writers such as Dahya (1974), Werbner (1984) and Helweg (1986), presenting these groups as pursuing dedicatedly economic goals within British society and remaining relatively aloof from the temptations of consumerism, leisure and the other customary fripperies of (post-) modern life. Consequently, even as low wage earners, they possess a high propensity to save, which obviously makes them very effective accumulators of capital to finance business entry.

While we are not entitled to read too much into Table 5.5, it must be said that it provides no immediate evidence for the last above-mentioned point, with Asians shown to be considerably less reliant on self-financing than either of the other groups. They are also less dependent on personal savings than on bank loans, with a substantial proportion of firms making no use of personal savings whatsoever. This does not entirely square with previous evidence: for example,

Table 5.5 Personal savings as start-up finance

Proportion of start up derived from personal savings (%)	Whites	Asians	Afro-Cbn	All
		(per cent of firms)		
None	48.1	46.1	40.4	46.1
Less than 25	9.3	12.2	2.4	9.8
25–75	5.4	16.7	14.3	11.9
Over 75	3.1	7.0	4.8	5.2
100	34.1	16.7	38.1	26.3

Rafiq (1985) finds personal savings to be the most frequent method of funding for Asian firms in Bradford. As in the present case, however, Rafiq finds that Asian firms (much as non-Asian firms) often use a mixture of sources, typically combining personal savings with bank credit in varying proportions. This is broadly echoed by McGoldrick and Reeve (1989) for Kirklees in Yorkshire, though these authors also find Asian firms relying exclusively on personal savings to be the largest single category, a pattern which is very definitely not replicated in the present instance. Indeed our own Asian respondents are less than half as likely as either of the other groups to be exclusively dependent on this method.

All this is mildly surprising in the light of all that has been said and written about the Asian propensity to work hard and save. A further special consideration which might be thought to strengthen still more their reliance on non-market finance is the possible influence of traditional religious beliefs, notably Islam, in which usury is absolutely prohibited. As we discovered in a previous survey of Bradford, *strict* Muslims will not give or take interest and more than one of our respondents told us that they managed to operate their firms successfully while abiding by this tenet (Jones *et al.* 1989). More usually, however, necessity forces adaptations not strictly within the letter of Islamic law and in practice this consideration is unlikely to apply to more than a small minority of Muslim firm owners (Bashir 1991). Certainly no Asian respondent in the present sample mentioned this as affecting their financial practices.

Evidently the relatively low personal savings input to Asian business needs alternative interpretation. One obvious hypothesis is that Asian financial behaviour is undergoing change with the passage of time, much as suggested by Deakins *et al.* (1992). Our own findings are immediately supportive of this suggestion, with those totally reliant on personal savings much more likely than others to be over 35 years old and born in the Indian sub-continent (Table 5.6). Only one of the Asian owners in this group is British-born and the younger generation is also emphatically under-represented, suggesting that a British upbringing is associated with a move away from traditionalist practices. Table 5.6 also hints at an association between educational attainment and use of formal financial institutions. Not ostensibly related to any of this, it is also shown (under the item 'positive business entrants') that members of this sub-sample are more likely to have been pulled into business by the lure of independence, money or some other positively defined goal than to have been pushed by unemployment or other negative labour market

Table 5.6 Personal characteristics of 26 Asian owners using 100 per cent personal savings for start up

Per cent who are	100% personal savings (I)	Rest of Asian sample II	¹/₁₁ × 100
Under 35	23.1	40.4	57.1
Female	12.5	17.4	71.8
Born in sub-continent	69.2	55.9	123.8
Children of business owners	44.0	43.4	101.4
With a degree	11.5	25.6	44.9
'Positively' motivated	65.3	55.9	117.0

conditions. Logically, of course, it is only to be expected that the stringent self-denial often required to accumulate self-finance needs to be supported by some definite expectation of personal gain, whether this be realistic or a form of self-delusion. When this is set against the other characteristics of this sub-sample, the suggestion is that the younger, more highly educated British-born members of the community are rather less likely to display such gritty stoicism. Far from a moral judgement, this is merely to note a gradual cultural mutation, which is bound to affect even members of such tightly enclosed communities as Asians in Britain.

Leaving aside differences in personal attributes, it is apparent that Asians in general are better placed than Afro-Caribbeans in respect of access to alternatives to personal savings. Whatever their unusual capacities for personal saving and whatever the advantages conferred by this interest-free low-risk mode, it would seem that commercial borrowing is now widely preferred. It would also seem that it is fairly widely available, contributing some proportion of start-up funding for three in five of the sampled Asian firms. While this certainly lends further weight to previous evidence that Asian entrepreneurs' credentials are now more likely to be recognized by bank lenders, later sections of this chapter will argue that their relationship with banks continues to be problematic in many respects.

Most dependent on self-finance are Afro-Caribbeans, a majority of whom raise one-quarter or more of their start up by this means, even though the Afro-Caribbean community is rarely if ever presented as possessing the kind of economic instrumentalism attributed to Asians. This lends weight to the argument that entrepreneurial strategies are less determined by migrancy or ethnic cultural predispositions than by sheer economic expediency under very tight constraints. As we shall see, the Afro-Caribbean relationship with banks tends to be even more problematic than is the case for Asians. Even at this stage it is clear that their unusually high dependence on personal savings is a corollary of this. Judging from the frequency with which our Afro-Caribbean respondents have encountered varying kinds of resistance from the high street banks, we are bound to conclude that this dependency is as much compelled as chosen.

In the interests of overall balance, we should not overlook the white sample, who also show a high propensity to use personal savings. Again this is in line with all that is known about standard small business practices in Britain, but which is usually written out of ethnic minority business studies. Additional weight has recently been lent to this point by Storey and Strange (1992), who find well over half of their (presumably mainly white) entrepreneurial sample mentioning personal savings as partly or wholly financing their start ups. Our own white owner sample seems to offer still further confirmation of self-financing as an established small business practice not in itself necessarily linked to ethnic origin.

FAMILY AND FRIENDS

Observers of successful ethnic minority business communities in the USA, notably Chinese and Japanese, lay considerable weight on the pooling of capital between fellow group members and the giving of interest-free loans on the basis of trust and common fellowship (Light 1972). As Wilson and Stanworth (1987) remind us, the great advantage of such non-market borrowing is that it is low cost, in effect giving ethnic minority entrepreneurs a competitive edge over mainstream firms, who presumably must borrow at commercial rates. Almost inevitably parallels have been drawn with Asians in Britain – though not with Afro-Caribbeans who are deemed to lack such in-group solidarity. Much of the literature on British Asians focuses on the traditional patriarchal extended family as a quasi-economic unit and a source of pooled funding. In the case of Pakistanis, for example, Werbner (1984: 167 and 169) depicts them as drawing on 'cultural resources which are perhaps unavailable to longer established minorities or the host society' and talks of members of the joint family contributing 'earnings to a common pool. Although each may earn very little, the total of accumulated earnings can be quite substantial.' The family is seen by its members as a 'joint enterprise having corporate aims', in effect a firm in its own right. Here is a clear case of the advantages to be derived from what Light and Bonacich (1988) call *acculturation lag*, the retention of a pre-modern institution within modern society.

Though we might have expected this family and group solidarity to have waned somewhat with the passage of time, commentators continue to stress the importance of family and friends as a source, with Asians predictably much more likely to have recourse to this than Afro-Caribbeans (Wilson and Stanworth 1987, Gretton 1988, McGoldrick and Reeve 1989, Deakins *et al.* 1992). For Soni *et al.* (1987: 81), the 'strong access of Asian entrepreneurs to capital from within the family' remains one of their strongest assets.

Quite clearly our own findings do not directly bear this out (Table 5.7). Contrary to standard expectations, it is the Afro-Caribbean sample who are most likely to be heavily dependent on family and friends, though the strength of this finding is somewhat obviated by the small Afro-Caribbean cell size. What is certainly incontrovertible is that both minority groups are more geared to this method than whites. Even so, it might be thought more significant that in none of the three cases does this form of funding make more than a minor contribution to the overall mix. Moreover there are hardly any individuals who have relied exclusively on this kind of sourcing. More often than not loans and gifts from family and friends are small and used to top up larger sums from elsewhere which make up the core of the starting capital. On this point we should note that several previous researchers, while finding that Asians use this source more than Afro-Caribbeans, have also found that Asian family and friends' input tends to be minor in comparison with bank loans and self-funding (Jones *et al.* 1989, McGoldrick and Reeve 1989).

While we are certainly not entitled to dismiss family and friends as a marginal contribution of little real account, it is nevertheless difficult to escape the feeling that perhaps their impact has been over-argued as part of a tendency to romanticize and idealize 'family', 'community' and other well-loved nostalgic reminders of a mythical vanished Golden Age. In the opinion of one prominent Asian business insider, there is a common misconception shared by bankers as well as the public at large that there is abundant communal wealth

Table 5.7 Start-up loan from family and/or friends

Per cent of start up	Whites	Asians	Afro-Cbn	All
		(per cent of firms)		
None	92.3	75.5	85.7	83.5
Less than 25	2.3	7.1	2.4	4.6
25 to 75	0.8	12.2	2.4	6.4
Over 75	0.8	0.0	0.0	0.3
100	3.8	5.1	9.5	5.2

to be shared within the Asian population. Though this may have been true twenty years ago, the Asian population is now 'too big and fractured' for such high intensity community spirit to survive (Bashir 1991).

Having administered this corrective we would nevertheless acknowledge that there may be other less direct ways in which family and ethnic community continue to exert an influence unrevealed by the evidence presented in our tables. According to Wilson and Stanworth (1987), established Asian business owners frequently act as loan guarantors for other family members, rather than furnishing a direct loan themselves. The present sample contains several Asian respondents who started off this way. Typical cases include a Sikh grocer in Gravesend, who raised his entire capital through a bank loan secured 'on my brother's track record'; and a young newsagent in Shrewsbury, whose business is essentially a spin-off from an established family concern, with the parent firm providing financial security for his loan, together with the reassurance that comes from a demonstrably sound record over several years of trading. We might once again see such cases as consistent with a cumulative developmental process, the Asian business community building initial strength and repute from a modest inward-looking self-sufficient base and progressively moving upwards and outwards into the commercial mainstream.

A further way in which the family might launch new business entry is through inheritance, where the heir is presented with a fully capitalized up-and-running outfit; or by opening branches, which may be capitalized by the parent firm and managed, sometimes with a considerable degree of autonomy, by members of the family. As can be appreciated, this is a relatively painless way to initiate the up and coming generation into entrepreneurship. At just over 13 per cent, the proportion of Asian respondents entering through inheritance or family acquisition is virtually identical to that of whites, whereas Afro-Caribbeans again appear disadvantaged, with as yet extremely restricted opportunities for acquiring ready-made firms from other members of the family. Of 54 Afro-Caribbean respondents, we could identify only a single instance of this: a young woman who had left a job in London to take over a deceased relative's grocery in Northampton. The rarity of this event underlines once more the recent emergence of an Afro-Caribbean business population in Britain and the consequent lack of accumulated resources.

Using the vocabulary of Light and Bonacich (1988), we would argue that the Asian business community in Britain has now reached

Table 5.8 Age of firms

Per cent established	White	Asian	Afro-Cbn	All
Less than five years	48.6	49.7	73.6	52.8
Five or more years	51.4	50.3	26.4	47.2

the stage ('critical mass'?) where it has recourse to a considerable stock of *class* resources: i.e. 'the money, human capital, materialistic values and business skills bourgeoisies normally possess' (Light and Bonacich 1988: 178). Partly this is a matter of growing maturity, the Asian business community being now long enough established to have built a sufficient stock of capital plus various non-monetary assets to give it a certain self- momentum (Ward 1991). To throw some light on this, Table 5.8 gives an age breakdown of firms, showing that the Asian sample contains an important core of fairly long established firms, though as might be expected this is not yet on a par with white firms. Reflecting their generally much later business emergence, the Afro-Caribbean firms sample shows a markedly more youthful age structure, implying that any cumulative momentum enjoyed by Asians is not yet shared by Afro-Caribbeans.

Yet the processes which underpin any Asian business momentum may be more complex than meets the eye. Certain obvious assumptions cannot be taken for granted, and in particular, we cannot blithely assume that inheritance will always occur as a kind of automatic cloning process by which Asian family firms are continually reproduced. An interview question was posed as to whether the respondent proposed to hand on the business to his/her children in the future and this yielded far from unanimous support for the idea. Certainly Asians were far more likely than the other two groups to be in favour of handing on their firm to their offspring (Table 5.9), with several of these affirmative replies expressed in such a way as to contemplate no real alternative. 'Keep it in the family', 'It's been set up for them', 'We want the kids to be established in business', 'Passing on to the children is the ultimate purpose of these ventures' are some of the responses that we might have expected if we viewed the Asian family entirely as a pre-modern patriarchal institution with a built-in sense of its own pre-ordained destiny. It clearly is not, however, because many other Asian business parents expressed vehement opposition to the notion of handing the firm on, often not wishing to condemn their heirs to what they saw as a life of unending toil for scant return, often because their children had more promising career options outside.

Table 5.9 Respondents hoping to pass business on

	White	Asian	Afro-Caribbean	All
Yes	28.6	60.5	33.3	44.0
No	71.4	39.5	66.7	56.0
Total per cent	100.0	100.0	100.0	100.0
Total count	119	129	27	275

Note: There are 135 missing observations, mostly either respondents without children under 18 or absent owners whose questionnaire was answered by a representative.

This last point echoes previous findings from the USA, where writers like Bonney (1975) and Bonacich and Modell (1980) stress that entrepreneurship has been more important as a launching pad for children into education and the professions than as a direct means of upward mobility in itself. If there is a prevalent attitude to this question, it is best captured by the clothing manufacturer in Coventry who insisted that, 'These days, parents do not force children into business'. There were many variations on this form of words, the eminently human response of parents the world over; and one which was indeed widely echoed by Afro-Caribbean and white respondents also. Whether this is a sign of creeping westernization is impossible to gauge on this evidence. Whatever the precise origins of these attitudes, their mere articulation reminds us yet again that common humanity should take analytical precedence over ethnic exoticism.

Returning to the theme of class resources, we should also remind ourselves that, in the Asian case, migration to Britain has not consisted purely of recruits to low grade occupations but has also contained an important minority with experience of business ownership in the homeland. In the 1970s, this was given an important boost with the influx of Asians from East Africa, where as is well known, Asians constituted a formidable business presence. While we are warned against the inevitable tendency to exaggerate this (Jamal 1976, Tinker 1977), it is nevertheless true that former business owners were over-represented among East African immigrants to Britain and, as we discovered in a previous study of Leicester (a major magnet for Ugandan refugees), many had somehow managed to bring money and other tangible assets with them (Aldrich *et al.* 1984; see also Soni *et al.* 1987).

Irrespective of the material resources furnished by family and community, it is possible that the true key to the reproduction and expansion of the Asian business community lies in less tangible forms of capital. The presence of Asian business role-models and precedents is crucial in building confidence and defining what is seen as

possible. Several young Asian respondents told us they had gone
into business 'to follow my father', to 'emulate my brother' or
because 'I saw a lot of other Asians doing this.' It is certainly the
case that Asian business leaders devote considerable energy to
building an image of their communities as 'naturally' predisposed to
successful entrepreneurship (Werbner 1990, Jones *et al.*
1992a, *Asian Business passim*) and, whether true or false, such a self-image is
almost bound to rub off on many members of the rising generation.
Because of this, young Asians can take themselves seriously as
potential entrepreneurs. It is of course the lack of such a self-image
that is assumed to stand in the way of Afro-Caribbean enterprise
development.

RELATIONS WITH THE HIGH STREET BANKS

In Britain, high street banks are a recognized source of start up and
working capital for small business. They are, however, not a particularly easy source, putting a high priority on security and business
track record, thus effectively placing new start applicants with few
tangible assets at a disadvantage (Woodcock 1986). From the bank's
own viewpoint this is a legitimate strategy of risk minimization, but
from a wider perspective it cannot but act as a dampener on good
business ideas. The fact that ethnic minority business applicants tend
to encounter even greater than normal resistance first came to light
in a number of early studies, notably that by Ward and Reeves
(1980), who showed that banks are not equal opportunity providers
of credit. In comparison with white entrepreneurs, ethnic minority
loan proposals are more likely to be rejected, compelling many
proposers to shop around extensively to obtain funding. Where credit
is available it is often at the cost of adverse terms, with banks
imposing excessive security conditions (Ward and Reeves 1980,
Brooks 1983, Wilson 1983, Reeves and Ward 1984, Wilson and
Stanworth 1987, Soni *et al.* 1987, Jones *et al.* 1989). As we have seen,
Asians and Afro-Caribbeans are now thought to be differentiated in
this respect also, a belief which is lent some support by the data
presented in Table 5.10. This shows Asian respondents as in some
respects more reliant on bank finance than whites, being far less
likely to be altogether without start-up bank loans and equally likely
to be exclusively bank-funded. It is the Afro-Caribbean sample who
stand out as most likely to be deprived of bank funding, either as a
result of having their credit applications refused or of avoidance on
their own part.

Table 5.10 Bank loans as start-up capital

Per cent of start up	Whites	Asians	Afro-Cbn	All
		(per cent of firms)		
None	55.8	38.7	61.9	48.5
Less than 25	3.9	5.2	0.0	4.0
25–75	6.2	20.0	16.7	14.1
Over 75	9.5	10.3	0.0	8.6
100	24.8	25.8	21.4	24.8

Further insight on this last point is given by Table 5.11, which summarizes responses to the question 'Did you have problems getting a bank loan?'. As the table immediately suggests, a very large proportion of those responding to the funding questions had not applied for any kind of start-up bank loan, either because this was deemed unnecessary or because of a wish to avoid anticipated problems of the kind discussed earlier. Quite predictably Afro-Caribbeans are the most frequent avoiders but, more unexpectedly, white respondents display a virtually identical frequency – unexpectedly, that is, if we choose to ignore the literature on the traditional economic behaviour of small business owners in countries like Britain and, in particular, recent findings on their financial methods (Storey and Strange 1992). It is the Asian sample who stand out as by far the most willing to plump for the bank option, a reflection presumably of a growing confidence within the Asian business community that they will now encounter a more welcoming reception than was formerly the case.

Judging by the problems reported in Table 5.11, however, this confidence is not always justified. Almost one in three Asian loan applicants claim to have encountered some kind of resistance, either leading to their application being turned down or to conditions being imposed which they thought were unjustifiably stringent. Yet Afro-Caribbeans have unquestionably suffered even more frequently, with almost two in five frustrated or hampered in their efforts to raise capital from the market. Once again we reach a point in the argument where it is essential to disentangle the specific circumstances pertaining

Table 5.11 Problems obtaining bank loans

	Whites	Asians	Afro-Cbn
Applicants	57.0	93.0	18.0
Applicants as % of respondents	44.2	60.0	42.8
Problems obtaining loan	12.0	27.0	7.0
Problems as % of applicants	21.0	29.0	38.9

only to racialized minorities from those applying to the general population of small business owners. It is not sufficient to simply state that the white position is less problematic than that of minorities, of whom Afro-Caribbeans are even more problematic than Asians. When the evidence from Table 5.11 is taken together with business owners' often jaundiced comments, we are bound to conclude that those dissatisfied with their bank transactions are too numerous to be ignored or written off as malcontents seeking a scapegoat for their own shortcomings. Nor are grievances solely confined to ethnic minority entrepreneurs, though these latter are of more than usual concern, because they occur more frequently and because their problems are usually self-diagnosed as stemming from racial bias on the part of the bankers.

While the white sample have certainly experienced problems less frequently than Asians and Afro-Caribbeans (Table 5.11), it is still the case that more than one in five applicants encountered some difficulty in securing a loan. Some of their comments are decidedly revealing. Not altogether unrepresentative is the Luton greengrocer who scornfully complained, 'Banks are no use. All the ads on telly are lies when you go to see the manager.' Doubtless encouraged by the banks' relentless self-promotion as the ultimate in user-friendliness, our disillusioned respondent appears to have stumbled upon 'The Inattentive Bank That Likes To Say "No!"'. This mood of disillusionment is detectable among many other white owners. It is part of an all-embracing demoralization: bolstered by official pronouncements in the 1980s that entrepreneurs were about to inherit the earth, they found that their own enterprise cut little ice with the guardians of the purse, private and public alike. Once again Luton, unfailing source of outspoken grassroots criticism, scores a bullseye, 'With this government, they trick you into business, then they put on a big business tax, poll tax and interest rates soar. It's all a con' (White clothing retailer).

Quite patently, then, membership of a white man's club counts for less than pure economic criteria in determining access to commercial credit. This is a generic factor appertaining to all small business and exacerbated at the time of the survey by the onset of recession. At the same time, ethnic *non*-membership can be critically disabling, imposing an additional and qualitatively different layer of resistance. This is well recognized in an editorial in the magazine *Asian Business* (7 June 1991), which diagnoses ethnic minority businesses as 'doubly excluded', subject to all the normal mechanisms of exclusion from banking–commercial networks and further marginalized by racially

specific mechanisms. Such feelings are widely shared by entrepreneurs on the ground, a selection of whose comments is reproduced
below.

Afro-Caribbeans

'I was refused by B*rc**y's but no satisfactory reason was given.
West Indians in business don't stand any chance.' Northampton
hairdresser.

'Banks are definitely not helpful to black people in small business'
Record shop proprietor, Northampton.

'Banks see West Indians as a bad risk.' Northampton restaurateur.

'I've approached banks several times but they don't want to know.'
Boutique owner, Lewisham.

'We need a black bank for black business.' (Lewisham grocer)

Asians

'We are disadvantaged because of colour.' Bexley CTN.

'When you go to the bank and the manager sees a black face, he
thinks twice.' Ealing takeaway food purveyor.

'My bank did not believe Asian people could join the export business'
Highly successful Preston exporter (of whom more anon).

As we noted in an earlier survey of unequal opportunities in business
(Jones *et al.* 1989), such comments should in a narrow technical sense
be regarded as suspicions rather than established fact. Only extremely rarely will individuals go to the lengths of proving discrimination, though we did find one extremely enlightening case in Bradford
in the earlier study (Jones *et al.* 1989). The present survey contains
a parallel case, an Afro-Caribbean entrepreneur in Northampton
who 'met with great resistance from the bank until I put my foot
down and accused the manager of race prejudice outright. They'll
give you anything to keep you quiet and get rid of you'. It is not
difficult to imagine the kind of personal courage required to mount
such a confrontation nor to empathize with the more usual inclination

of intimidated clients to keep their suspicions to themselves. Even so, suspicions are so widespread that they can hardly be without foundation and in any case their mere existence must undeniably be highly damaging to the confidence of entrepreneurs and to communication between bank and customer. We find nothing in the testimony of our present respondents to overturn the weight of evidence accumulated by a string of writers from Ward and Reeves (1980) to the present.

Having stated this, however, we must underline that witch-hunts are not in order: and that to point a moralizing finger at bank managers as an inherently racist set of individuals is singularly counter-productive as well as scientifically suspect. As Ward and Reeves (1980) are at pains to point out, their ethnic minority respondents found a great deal of variation in individual attitudes on the part of bank managers and other officials. In the interests of balance, we note that this is once more replicated in the present survey, with significant numbers of individuals positively complementing their banks for helpful treatment. Unhappily, they are outnumbered among those who ventured comments by those who are critical and suspicious, more so for Afro-Caribbeans than Asians.

Doubtless, in rare cases individual bank managers and officials may be quite consciously racially biased, though this does not mean that they will necessarily translate their prejudices into practice. Moreover, if they do, this is likely to occur in a covert fashion, so that the victim will have no *tangible* grounds for grievance: what Sivanandan (1978) calls the 'genteel racism of the middle class' as opposed to its more blatant working-class variant. Much more generally, however, we would argue that the process is likely to be subconscious or subliminal. By this we mean that certain negative images of former colonial peoples are so deeply embedded in the European psyche that even consciously well-intentioned individuals may be infected with a stereotypical mindset unconsciously acquired, taken in with their mother's milk so to speak. Bankers are as much products of this cultural socialization as anyone else and consequently with the best will in the world it may be difficult for them to see Afro-Caribbean or Asian people as credible in the role of entrepreneur.

None of this is to condone or to condemn, but merely to attempt to arrive at a non-moralistic understanding. On this basis it would seem that there are some fairly obvious steps banks could take to rectify matters. They could work towards heightened awareness on the part of white staff and they could ensure that ethnic minority staff were placed in key decision-making positions, thereby alleviating problems of non-communication and alienation on the part of black customers. We welcome recent official pronouncements by the major high street banks that they do indeed recognize the problem and are taking the kinds of steps envisaged above. Even so, the feedback from our respondents indicates that there is still a considerable gap between official head office policy and concrete practice at branch level.

INTRA-ASIAN VARIATIONS

One question still to be resolved is the apparent contrast in the way Afro-Caribbeans and Asians are treated at the bank, with bank managers seemingly prepared to make a distinction, albeit limited, in favour of the latter. This view of events needs careful clarification. Even if, as is the case with our respondents, Asians complain less frequently about racial discrimination, any complaint whatsoever is cause for concern and we are certainly not entitled to the complacent conclusion that Asians now enjoy an easy non-problematic relationship with the banking system. Table 5.11 makes it quite clear that there is a substantial proportion of Asians who feel in some way frustrated and uncomprehending at their treatment, with a lurking suspicion that, precisely because they *are* Asian, they are not being treated like other business loan applicants.

The truth of the matter is that there are not simply differences between Asian and Afro-Caribbean loan applicants but also important differences between some Asian entrepreneurs and others. Whether by accident or design, bank investment in Asian business is in practice highly selective, as is demonstrated by Table 5.12. Here we have isolated the forty Asian firms who obtained the whole of their start-up funding from banks, and the table compares this subsample with the total Asian sample along a range of variables, whose over- or under-representation is given in Column III. For us, the salient feature here is the *sectoral* composition of the sub-sample. Those who were 100 per cent bank-funded contain a marked over-representation of low order retail firms, notably those in Confectionery-Tobacco-News (CTN). Among the inferences to be drawn, the most obvious is that firms in activities which have become typecast as 'Asian' – i.e. branches of enterprise like CTN and grocery shops, in which Asians are known to be highly concentrated and thus assumed to be viable and secure – enjoy easier access to commercial funding than those in 'atypical' sectors where presumably there is a higher degree of perceived risk.

Table 5.12 Profile of 40 Asian owners obtaining 100 per cent start up from bank loan

	40 owners (I) %	Rest of Asian sample (II) %	$^1/_{11}$ × 100
Personal			
1 Under 35 years	45.0	39.1	115.1
2 Male	80.0	81.1	98.6
3 Self-employed parent(s)	30.0	47.1	63.7
4 British-born	15.0	9.4	159.6
5 University/college degree	27.5	22.4	122.8
6 Entry motive positive	58.0	58.7	99.3
Firm			
7 5 years old or less	46.0	51.4	89.5
8 Food retail + CTN	58.0	29.7	195.3
9 Manufacture and W/sale	0.0	3.5	0.0
10 Premises owned	57.5	53.6	107.3
11 Premises problems	27.5	27.5	100.0
Practice and performance			
12 Profits unsatisfactory	32.5	26.8	121.3
13 Personal income unsatisf'y	27.5	21.7	126.7
14 Owner's workload 80+hrs/wk	37.5	15.9	235.8
15 Employing 10+ FT workers	2.5	10.1	24.7
16 Using unpaid workers	52.5	21.0	250.0

Quite clearly, then, bank investment in Asian small business is not only rather modest *in toto* (some would say completely inadequate) but also unimaginatively targeted at a very limited range of firms, thereby reinforcing the narrow sectoral entrapment of the Asian business economy. Unsurprisingly, annoyance has been expressed from within the Asian business community at this stereotyping of its members as 'corner shop operators' (Bashir 1991). In a back-handed way, such stereotyping may be in the narrow short-term interests of low order retailers needing bank finance, but it is not conducive to the wider long-term development of Asian enterprise *per se*.

A further inescapable suggestion from Table 5.12 is that these branches include a heavy concentration of the most marginal firms, smallest in scale and surviving on the back of heavy entrepreneurial workloads and uncosted labour (items 14–16). Not unrelated to this, low earnings firms are also prominent here (items 12–13). Unhappily for those who prefer tidy packaging, it is the least modern and formal of firms who are the most likely to use the formal credit market, a somewhat contradictory juxtaposition. Whatever we make of this, it is evident that banks are making little impact on a major weakness in the Asian business economy, its acute constriction into a narrow band of easy-to-enter but low margin labour-intensive branches of activity (Jones 1983, Jones *et al.* 1989, 1992c, Jones and McEvoy 1991; see also Rafiq 1985, McGoldrick and Reeve 1989).

Earlier in the chapter it was observed that one of the hallmarks of ethnic minority business is the substitution of labour for capital. Elsewhere, we have also established that Asian entrepreneurial earnings tend to be better in those firms and branches where the owner's workload is lighter, i.e. where money capital or human capital in the form of qualifications and expertise has been substituted for simple toil (Jones *et al.* 1992c). Judging from Table 5.12, however, there is only minimal bank investment in Asians who are breaking out of the sectoral strait-jacket into less crowded areas like manufacturing and wholesaling. From Table 5.12 it can be inferred that such ventures tend to be higher earning, larger in scale and, though still labour-intensive, imposing less gruelling workloads on their owners. Perversely of course these branches have higher entry thresholds, requiring much more expertise and, above all, capital than is the case for low order retailing. With heavy irony, it is the very branches where bank funding could be crucial which are the least likely to attract it.

From the bankers' own perspective it might be argued that failure to back such businesses represents a host of missed opportunities to become involved in profitable and expansionary Asian enterprise development. Several of the most successful Asian ventures in the sample have achieved this entirely without bank investment or with only marginal backing. One outstanding instance is an East African entrepreneur, who now has an export company employing twenty-five workers in Preston together with three African branches. In this he was able to confound his bank manager who 'did not believe Asian people could join the export trade'. Join it with a vengeance our respondent did, and he is now one of the rare individuals who can declare himself better than satisfied with profits and income, and who has expanded consistently over a decade.

Among this category who had to rely mainly or wholly on non-bank sourcing, there is also a substantial concentration of manufacturers, who are completely absent from the forty owner sub-sample. Of the twenty Asian manufacturers interviewed, four claimed to have had their proposals turned down by the bank. The rest are presumably abstainers or partial abstainers who have adopted alternative strategies to by-pass the banking system, possibly in anticipation of delays, costliness and frustration, where avoidance represents the line of least resistance.

On the whole, this must be seen as a loss. Although the benefits of Asian manufacturing defy easy evaluation (notably in their impact on Asian workers: see Mitter 1986 and Ram 1990 for discussion), it is fair to say that from the entrepreneur's viewpoint, entry into manufacturing represents a marked advance in scale and content from the stereotypical corner shop. In Britain, manufacturing sectors like clothing, textiles and knitwear have been deeply penetrated by Asian ownership, particularly in localities like Leicester (Soni *et al.* 1987). In so far as the present sample can be called nationally representative, however, it suggests that generally manufacturing is still a minority pursuit. Less than 4 per cent of Asian respondents are in manufacturing compared with around half in the three retail branches of food, CTN and clothing. In the present survey, Asian manufacturers were overwhelmingly located in Preston and Coventry, though here we remind the reader that this is partially a reflection of the non-random selection of localities and of small areas within them.

Most commonly, Asian manufacturers cited their own personal savings as their principal or sole source. In practice this is frequently matched or exceeded by *supplier credit*. More often than not, clothing manufacturers were contracted to sell the whole of their output to fellow Asian wholesalers and, although we cannot exhaustively quantify this, we are fairly certain that it provides an important avenue for start up. In effect, this funding method combines a formal commercial relationship with all the informal advantages of an ethnic insider network. A principal authority on this subject is Werbner (1984, 1990) who demonstrates that credit arrangements within the Asian clothing network are highly liberal and ethnic-preferential. She quotes a Manchester Pakistani wholesaler to the effect that 'Asian traders expect credit as a right' (Werbner 1984). According to Werbner, Manchester lies at the centre of an extensive and complex web of Asian rag trade entrepreneurs and it is evident that our own survey has picked up many outliers of this web; manufacturers and wholesalers in Preston, retail stall-holders in Liverpool. Taking advantage of the arrangements described, a jeans manufacturer interviewed in Preston had been able to float his business with £20,000 credit from his supplier and on the basis of this and several similar instances, we are able to approximately confirm the importance of supplier credit for Asian manufacturers. We would also presume that, because of the insignificant number of Afro-Caribbean suppliers, this option is virtually non-existent for entrepreneurs in that community. Certainly in the case of our own Afro-Caribbean firms sample, there was only marginal input from co-ethnic suppliers.

Once again, however, we should not be seduced by the temptations of ethnic particularism: according to Deakins *et al.* (1992), 'Trade credit remains an important source of short-term finance . . . [but] . . . *this is little different from non-ethnic businesses*' (our italics).

As well as being distinguished by sector, scale and performance, sub-sample presented in Table 5.12 differs in several other respects also. Personal characteristics and background (items 1–6) would be expected to exert some influence on financial behaviour and here we find that British-born and more youthful respondents are over-represented, strongly in the former case. This may well indicate that dependence on informal ethnic linkages is tending to wane over time (a suggestion confirmed by Deakins *et al.* 1992), with the rising generation of entrepreneurs less hampered by linguistic and psychological barriers and hence more willing and able to access the banks. Presumably such a trend will be welcomed and actively cultivated by the banks themselves. However, any sanguine optimism about the benign influence of the passage of time is tempered by the finding that the most recent wave of firms are no more likely to gain access to bank backing than the longer established firms (item 7).

A final noteworthy feature is contained in Table 5.12, item (3), showing that offspring of business families are considerably less likely to use banks than are Asians as a whole. Doubtless this reiterates the importance of the family firm as a direct source of capital for spin-off enterprises. In principle it might be imagined that a business family background would equip a budding entrepreneur with the knowledge needed to present a case for a bank loan, but this is evidently not a major force here.

AFRO-CARIBBEAN VARIATIONS

Of the forty-two Afro-Caribbeans who answered the questions on funding, only nine were exclusively funded by banks at start up. This compares with three times that number who obtained no bank backing whatsoever, in itself additional commentary on the great gulf yawning between these entrepreneurs and the banking system; and making it easy to understand the Northampton café-owner who said simply, 'I am in a white country trying to run a black business and it's hard'. Banks all too clearly symbolize that 'white country'.

Although a sub-sample of nine is technically of no statistical significance, there are nevertheless two features of the exclusively bank-funded group that are of more general significance: *eight* of them gave positive reasons for business entry – the need for independence, the urge to exploit a business opportunity or a skill – but only *one* was able to report satisfactory earnings. Since none of them gave 'to make money' as a motive, we can infer that their pecuniary expectations were rather modest in any case but that even these have not been met. Elsewhere we have drawn attention to a recurrent mismatch between the positive motivations of Afro-Caribbean entrepreneurs (a finding directly contradicting prevalent stereotypes) and their negative performance, suggesting that we should look for explanations beyond the entrepreneurs themselves and beyond the alleged deficiencies of their ethnic communities (Jones *et al.* 1992a). Here again we see a group of people who have come into business with good ideas and defined goals, who have succeeded in jumping through the flaming hoops posed by the banking system, but who are clearly not viable despite this.

Frequently this was attributable to external factors such as an impoverished inadequate customer base and to a surfeit of small firms vying for it – 'too much competition from Indian restaurants' as one owner put it. There were also mentions of crippling costs, as with the Northampton record shop owner who said, 'The only thing the council gives me is problems like rents, poll tax, business rates.' There were also specific mentions of high bank charges and interest rates, suggesting that the small owner is caught on the horns of a chronic dilemma *vis-à-vis* the banking system. Successful penetration of the banking system may furnish adequate capital, but only at a burdensome cost, while exclusion may mean underfunding. The entire question of under-funding will be confronted in the final section of this chapter.

POST-START UP FUNDING

It goes without saying that obtaining capital to start a firm, while a crucial hurdle in itself, is not the end of the matter. There is a recurring need for working capital and sometimes for finance for improvements and expansion. It must be said that our own respondents have more often than not met with fewer problems in obtaining bank loans at this stage, a finding which gels with the established wisdom that increasing maturity eases the problem of funding. Most pertinently, many entrepreneurs who were refused or who abstained from start-up loans were successful in mobilizing subsequent loans, usually for working capital. One such individual is the Preston exporter cited previously, a clear example of the effectiveness of a proven track record and a firm's tangible assets to provide security.

We presume that this principle must apply fairly widely to many other cases, even where not so outstandingly successful. Moreover, in the case of ethnic minority post-start up applicants, it is likely that managers are now able to judge them on individual characteristics and achievements, and thus may feel less compelled to fall back on stereotypical evaluation.

Almost inevitably, the picture presented in Table 5.13 is not entirely unclouded. Access to post-start funding is far from universal for any of the three groups. Asians are slightly better placed than Afro-Caribbeans, further possible evidence for their growing credit-worthiness. Several Asian manufacturers, having launched themselves in the complete absence of bank credit, have successfully resorted to it for working capital and expansion. While this can only be seen as positive in itself, it is nevertheless difficult to avoid a feeling that banks enjoy an enviable freedom to have their cake and eat it: avoiding any of the risks of launching a venture, they are able to make profitable investment after it is seaworthy and afloat. Such criticism is entirely futile of course, since commercial banks do not exist to provide venture capital. Along with others such as Mason *et al.* (1991), we wonder who does.

There still remains a sizeable proportion of firms (43 per cent) not using post-start bank credit. These include a fair number of very recently established concerns for whom the question has not arisen. It also includes many whose main source of working capital is profits ploughed back. Yet another time-honoured custom and practice, plough-back offers security, cost-saving and all the other advantages of self-finance, together with various potential disadvantages, such as under-funding and diminished current income.

Among these non-users of post-start up credit, Afro-Caribbeans stand out as substantially over-represented. Though this might be partly attributed to the large proportion of recently started Afro-Caribbean firms, there is also a suggestion that even longer established

Table 5.13 Main sources of post-start up funding

	White	Asian	Afro-Cbn	All
Per cent using post-start up funding	58.5	72.4	55.5	64.4
Of which:				
Bank	63.0	55.8	40.0	56.7
Personal savings	9.0	9.3	10.0	9.2
Profits	39.0	37.2	26.7	36.7
Family & friends	7.0	18.6	16.6	13.9

Note: Many firms used more than one method of funding

owners continue to encounter resistance or are discouraged from making overtures to the bank. Since Table 5.13 also shows that Afro-Caribbeans are less likely to be able to use personal savings, the implication is that Afro-Caribbean firms may well be seriously under-funded in respect of working capital as well as start-up capital.

AN INTERIM VERDICT

In this chapter, we have provided a substantial measure of support for much previous research into ethnic minority enterprise funding in Britain. Of all the findings presented here, we would highlight *racial disadvantage* in securing capital, a bias which can only act to inhibit the formation of ethnic minority firms through outright capital starvation, and to undermine the quality of those who do come into existence through financial under-nourishment. Principally emphasized has been the role of banks, whose lending practices (as opposed to *declared policies*) lead in effect to a misallocation of resources. On this we concur heartily with researchers such as Ward and Reeves (1980) and Wilson and Stanworth (1987), who have identified this as one of the most insurmountable barriers to black entrepreneurial advance.

We reiterate that this is not intended as a pious 'politically correct' attack on banks or their personnel. Racial bias there most certainly is, but this is best regarded as the unintended consequence of cultural conditioning, as indeed it is in most other spheres of economic and social life. Yet consequences are consequences, whether intentional or not. Banks are one of the circuits in a multiple cycle of disadvantage 'in which characteristics are attributed to the disadvantaged which then become the justification for their disadvantaged position in society' (Lee and Loveridge 1987). Other circuits are located in the labour market and in the various agents other than the banking system which control the allocation of entrepreneurial resources. Ram (1992) talks of racism as 'a constant and constraining feature in the environment that these [Asian] firms operate in', specifically citing the banking system as a critical component of this environment.

Elsewhere, we part company with some of the other authors, though sometimes this is a matter of emphasis or more often *de*-emphasis. Questions are raised in particular about many of the standard suppositions about Asian financial practices, where it seems that reports of their special propensity for individual and collective self-financing are either exaggerated, out of date, or do not take full account of some of the subtleties. Afro-Caribbeans are once again shown to be exceptionally disadvantaged in the matter of finance, more so than Asians, but this does not mean that Asians enjoy full and easy access to funding. On the contrary, their relationship to the high street banks remains in many respects highly problematic. From another perspective, it might be said that the banks are failing to draw out the entrepreneurial potential of either ethnic minority community.

Inclusion of a large contingent of white business owners also provides a somewhat different perspective from those studies which view ethnic minority enterprise in isolation. Certain forms of business behaviour which are widely thought to be specific to certain ethnic cultures are revealed to be broadly common to all small firms, as also are various problems of capital formation. This needs to be clearly recognized, but in so doing we should not for one moment assume that ethnic minorities do not suffer additional and special barriers. Much research remains to be done in this field. Bank managers and other key officials need to be more thoroughly researched to discover how they view their ethnic minority clients and to explore ways of heightening their awareness of ethnic minority special needs.

Finally we would acknowledge a deficiency. In common with most authors in this field, our research at this stage is decidedly biased towards the demand side of the financial equation (Deakins *et al.* 1992). Representatives of the supply side – most notably the bankers themselves – have as yet been given no voice. Given their absolutely critical decision-making role, this is a serious imbalance and it is clearly desirable that future research focus on bankers as subjects in their own right. No balanced conclusions can be drawn without a speaking part for all the major players on this particular stage. Arguably the bankers should be cast in the leading role.

NOTE

1 Asian interviews from Liverpool, Shropshire-Herefordshire and selected wards in Batley, Preston, Wolverhampton, Coventry, Swindon, Gravesham, Bexley, Ealing and Luton. Afro-Caribbean interviews from selected wards of Manchester, Nottingham, Northampton and Lewisham. White interviews in all fifteen localities.

REFERENCES

ACOST (1990) *The Enterprise Challenge: Overcoming Barriers to Growth in Small Firms*, HMSO, London.
Aldrich, H., Cater, J., Jones, T. and McEvoy, D. (1981) 'Business

development and self-segregation: Asian enterprise in three British cities', in Peach, C., Robinson, V. and Smith, S. (eds) *Ethnic Segregation in Cities*, Croom Helm, London.

Aldrich, H., Jones, T. and McEvoy, D. (1984) 'Ethnic advantage and minority business development' in Ward, R. and Jenkins, R. (eds) *Ethnic Communities in Business*, Cambridge University Press, Cambridge.

Aldrich, H., Zimmer, C. and Jones, T. (1986) 'Small business still speaks with the same voice: a replication of the "voice of small business and the politics of survival"' *Sociological Review* 34, 57–88.

Asian Business (1991) 'Nazmu Virani: Champion of the Community', 13–26 September 1991.

Bains, H. (1988) 'Southall youth: an old-fashioned story' in Cohen, P. and Bains, H. (eds) *Multi-Racist Britain*, Macmillan, London.

Ballard, R. and Ballard, C. (1977) 'The Sikhs: The development of South Asian settlement in Britain' in Watson, J. (ed.) *Between Two Cultures*, Blackwell, Oxford.

Bashir, A. (1991) 'Tackling the barriers to enterprise', Conference on Entrepreneurs and Equal Opportunity, University of Bradford.

Bechhofer, F., Elliott, B., Rushforth, M. and Bland, R. (1974) 'The petite bourgeoisie in the class structure' in Parkin, F. (ed.) *The Social Analysis of Class Structure*, Tavistock, London.

Bechhofer, F. and Elliott, B. (1978) 'The voice of small business and the politics of survival' *Sociological Review* 26, 57–88.

Bonacich, E. (1973) 'A theory of middleman minorities' *American Sociological Review* 37, 547–59.

Bonacich, E. and Modell, J. (1980) *The Economic Basis of Ethnic Solidarity*, University of California Press, Berkeley.

Bonney, N. (1975) 'Black capitalism and the development of the ghettos in the USA', *New Community* 4, 1–10.

Brooks, A. (1983) 'Black business in Lambeth' *New Community* 11, 42–54.

Brown, C. (1984) *Black and White Britain*, Heinemann, London.

Cashmore, E. (1991) 'Flying business class: Britain's new ethnic elite' *New Community* 17, 347–58.

Creed, R. and Ward, R. (1987) *Black Business Enterprise in Wales*, South Glamorgan CRE.

Curran, J. (1986) 'The survival of the Petit Bourgeoisie: production and reproduction' in Curran, J., Stanworth, J. and Watkins, D. (eds) *The Survival of the Small Firm*, vol. II, Gower, Aldershot.

Curran, J. and Burrows, R. (1988) 'Ethnicity and enterprise: a national profile', 11th Small Firms Policy and Research Conference, Cardiff.

Dahya, B. (1974) 'Pakistani ethnicity in industrial cities in England' in Cohen, A. (ed.) *Urban Ethnicity*, Tavistock, London.

Daily Express (1990) 'Advice on Making Millions', p. 30, June 27.

Deakins, D., Hussain, G. and Ram, M. (1992) *Finance of Ethnic Minority Small Businesses*, University of Central England in Birmingham.

Gretton, E. (1988) *Black Business Development in Nottinghamshire*, Fullemploy and Nottinghamshire County Council.

Helweg, A. (1986) *Sikhs in England*, Oxford University Press, London.

Hiro, D. (1971) *Black British: White British*, Eyre and Spottiswoode, London.

Home Office (1991) *Ethnic Minority Business Initiative: Banks Report on Ethnic Minority Business Development*.

Jamal, V. (1976) 'Asians in Uganda 1880–1972: inequality and expulsion', *Economic History Review* 29, 602–16.

Jennings, K. (1991) *Observer*, March 17.

Jones, T. (1983) 'Residential segregation and ethnic autonomy' *New Community*, 13, 10–22.

Jones, T. and McEvoy, D. (1991) 'Ressources Ethniques et égalités des chances: les entreprises indo-pakistanaises en Grande Bretagne et au Canada, *Revue Européenne des Migrations Internationales* 8, 107–26.

Jones, T., Cater, J., De Silva, P. and McEvoy, D. (1989) *Ethnic Business and Community Needs*, Report to the Commission for Racial Equality, Liverpool Polytechnic.

Jones, T., McEvoy, D. and Barrett, G. (1992a) 'Ethnic identity and entrepreneurial predisposition: business entry motives of Asians, Afro-Caribbeans and Whites', Paper to the ESRC Small Business Initiative, University of Warwick, February 4.

Jones, T., McEvoy, D. and Barrett, G. (1992b) *Small Business Initiative Ethnic Minority Business component*, ESRC.

Jones, T., McEvoy, D. and Barrett, G. (1992c) 'Labour-intensive practices in the ethnic minority firm', Paper to ESRC Small Business Initiative, University of Warwick, September 2.

Lee, G. and Loveridge, R. (1987) 'Preface' to Lee and Loveridge (eds) *The Manufacture of Disadvantage: Stigma and Social Closure*, Open University Press, Milton Keynes.

Light, I. (1972) *Ethnic Enterprise in America*, University of California Press, Berkeley.

Light, I. and Bonacich, E. (1988) *Immigrant Entrepreneurs*, University of California Press, Berkeley.

McGoldrick, C. and Reeve, D. (1989) *Black Businesses in Kirklees*, Department of Geographical Sciences, Huddersfield Polytechnic, Huddersfield.

Mail on Sunday (1992) 'Minorities and the millionaires', July 5.

Mason, C., Harrison, C. and Chaloner, J. (1991) 'Informal risk capital in the UK: a study of investor characteristics, investment preferences and investment decision-making', Paper to ESRC Small Business Initiative, University of Warwick.

Mitter, S. (1986) 'Industrial restructuring and manufacturing homework', *Capital and Class* 27, 37–80.

Ohri, S. and Faruqi, S. (1988) 'Racism, employment and unemployment', in Bhat, A., Carr-Hill, R. and Ohri, S. (eds) *Britain's Black Population: A New Perspective*, Gower, Aldershot.

Patel, S. (1988) 'Insurance and ethnic community business' *New Community* 15, 79–89.

Patel, S. (1991) 'Patterns of growth: Asian retailing in inner and outer areas of Birmingham' in Vertovec, S. (ed.) *Aspects of the South Asian Diaspora*, Oxford University Press, Delhi.

Patterson, S. (1969) *Immigration and Race Relations in Britain 1960–67*, Oxford University Press, London.

Phizacklea, A. and Miles, R. (1980) *Labour and Racism*, Routledge, London.

180   *Trevor Jones, David McEvoy and Giles Barrett*

Rafiq, M. (1985) *Asian Businesses in Bradford: Profile and Prospect*, Bradford Metropolitan Council, Bradford.
Ram, M. (1990) 'Control and autonomy in small firms: the case of the West Midlands clothing industry' *Work, Economy and Society* 5, 601–19.
Ram, M. (1992) 'Coping with racism: Asian employers in the inner city' *Work, Employment and Society* 6, 601–18.
Ram, M. and Sparrow, J. (1992) *Research on the Needs of the Asian Business Community in Wolverhampton*, Wolverhampton TEC Ltd and Wolverhampton Chamber of Commerce.
Reeves, F. and Ward, R. (1984) 'West Indian businesses in Britain' in Ward, R. and Jenkins, R. (eds) *Ethnic Communities in Business*, Cambridge University Press, Cambridge.
Lord Scarman (1981) *The Brixton Disorders 10–12th, April 1981*, HMSO, London.
Scase, R. and Goffee, R. (1980) *The Real World of the Small Business Owner*, Croom Helm, London.
Shah, H. (1991) 'Labour urges action to help depositors' *Guardian*, 8 July.
Sivanandan, A. (1978) *A Different Hunger*, Pluto, London.
Smith, D. J. (1976) *The Facts of Racial Disadvantage*, Political and Economic Planning, London.
Soar, S. (1991) 'Business development strategies' *TECs and Ethnic Minorities Conference Report*, Home Office Ethnic Minority Business Initiative, Warwick University.
Soni, S., Tricker, M. and Ward, R. (1987) *Ethnic Minority Business in Leicestershire*, Aston University, Birmingham.
Storey, D. and Strange, A. (1992) *Entrepreneurship in Cleveland 1979–1989: A Study of the Effects of the Enterprise Culture*, Centre for Small and Medium Sized Enterprises, Warwick Business School, University of Warwick.
Tinker, H. (1977) *The Banyan Tree: Overseas Emigrants from India, Pakistan and Bangladesh*, Oxford University Press, Oxford.
Today (1990) '300 Asians Make A Million', August 25.
Ward, R. (1991) 'Economic development and ethnic business' in Curran, J. and Blackburn, R. (eds) *Paths for Enterprise: The Future of Small Business*, Routledge, London.
Ward, R. and Reeves, F. (1980) *West Indians in Business in Britain*, HMSO, London.
Weber, M. (1976) *The Protestant Ethic and the Spirit of Capitalism*, Allen & Unwin, London.
Werbner, P. (1984) 'Business on trust: Pakistani entrepreneurship in the Manchester garment trade' in Ward, R. and Reeves, F. (eds) *Ethnic Communities in Business*, Cambridge University Press, Cambridge.
Werbner, P. (1990) 'Renewing an industrial past: British Pakistani entrepreneurship in Manchester', *Migration* 8, 17–41.
Wilson, P. (1983) *Black Business Enterprise in Brent*, Runnymede Trust, London.
Wilson, P. and Stanworth, J. (1987) 'The social and economic factors in the development of small black minority firms: Asian and Afro-Caribbean

business in Brent, 1982 and 1984 compared' in O'Neill, K., Bhambri, R., Faulkner, T. and Cannon, T. (eds) *Small Business Development: Some Current Issues*, Avebury, Aldershot.

Woodcock, C. (1986) 'The financial and capital environment of the small firm' in Curran, J., Stanworth, J. and Watkins, D. (eds) *The Survival of the Small Firm* vol. I, Gower, Aldershot.

6 The use of management accounting information for managing micro businesses

Amanda Nayak and Sheila Greenfield

INTRODUCTION

The last decade has seen a rapid growth in the number of small businesses (Daly 1991) and a corresponding expansion in the amount of information and resources available to those wishing to enter self-employment. Small businesses have thus become a highly topical area for research. Curran (1986) however highlights the fact that research has tended to concentrate on larger-sized enterprises and neglect very small businesses run by one or two people with few or no employees (micro businesses) He stresses the need in particular for in-depth qualitative research on the one-person small enterprise. A major focus in small business research has been the characteristics of owner-managers (e.g. The Bolton Report 1971, Stanworth and Gray 1991). Much of the writing on how they run their businesses has been by way of case studies. There is a perceived view within the business community that these micro businesses do not keep adequate records. It was felt that a survey which looked at a large number of small firms might reveal interesting insights on how they gather, record and use the information which they need to run their businesses.

The study described in this chapter therefore set out to identify the management information available to and used by proprietors of micro firms in the day to day running of their businesses. Micro businesses with only one major decision maker are in a unique situation as only at this size can a firm manage without having to communicate information to other people within the business in order to work effectively. Small businesses can vary in size from substantial organizations of up to two hundred employees to one-man bands (for an excellent discussion see Curran 1989). The purpose of this study was to discover whether very small businesses which do not have to collect information for dissemination to other managers had enough information to pursue their goals. It was necessary, therefore, to research businesses with only one decision maker, and firms with under ten employees were an appropriate size for this purpose.

Businesses and management information

Business organizations have two main sources of management information which must be obtained: external information, obtained from outside the firm (e.g. concerning suppliers, business contacts, customers and changes in the business climate) and internal information generated by the business, which can be compiled from the activities in which it is engaged. Large firms need to collect information on a regular basis for organizing, communicating with and monitoring staff. As stated earlier, in very small businesses where there is only one decision maker, information does not have to be disseminated for these purposes. It is therefore difficult to discover how small business proprietors operate, as the information they use may not be collected in a formal way, but may be stored in the businessman's head. Some information which it might be expected that all businesses would have, may not have been acquired at all by some small businesses. Little has been documented about the information which micro businesses have available to help them to manage and exercise financial control.

The annual profit and loss accounts indicate, after the year end, how well a business has succeeded during the year. However, in order to achieve profit, businessmen must make successful decisions on the day to day running of the firm, long-term strategy and investment decisions. To do this they must have regular information. Accountants who regularly prepare the accounts of micro businesses are familiar with the 'shoe box' or 'spike' of paid invoices and the untotalled cash book with which they are regularly presented by some small businesses. Larger firms have formal recording procedures laid down for purposes of information flow between participants and internal control requirements. The fact that micro businesses are less organised in their record-keeping does not necessarily indicate a lack of organization and effectiveness, but may be due to a lack of need.

Existing research

Several studies have been undertaken to determine what financial ratios are used by owners/managers of small businesses to monitor their business (Lewis and Toon 1986, Holmes and Nicholls 1989, McMahon and Davies 1991). Attempts have been made in some cases to measure business success as a function of knowledge and use of these ratios. However, these studies have been on larger small businesses (i.e. more than ten employees) or growing small businesses. Other studies have been conducted into the role of the professional adviser as perceived by both adviser and client (e.g. Lewis and Toon 1986, Robertson 1988). Apart from a small study on similar issues which was conducted in 1985 by Wootton and Templeman, documentation on how firms are organised to use accounting information when they are very small (under ten employees) and how they obtain this information has not been extensively researched. Very small firms are unlikely to employ a full-time accountant or even, in some cases a book-keeper. Therefore the way they obtain managerial accounting information is likely to depend on their own activities and needs.

Carsberg *et al.* (1985) have suggested that an important function of company audit, even for small companies, is that it forces them to keep accounting records, thus implying that without these records companies are more likely to fail. Hankinson (1983, 1985) has done a series of in-depth studies of pricing in various different sectors and sizes of small firms.

UNDERLYING ASPECTS OF MANAGEMENT THEORY

Management theory defines the functions of management as:

planning
controlling
organizing
communicating and
motivating (Drucker 1974).

In order to perform these functions managers must have information. This is generated partly internally by keeping records of the activities of the firm and its relationships with outside parties and partly by searching the environment in which it operates for information on, for example, suppliers, customers, competitors, legislation, etc. The extent to which small businesses need this information and are able to collect and use it is largely unknown.

Planning and business goals

The function of planning implies that goals or objectives have been defined (since it is not possible to plan without a goal).[1] For the purpose of the current study there are two reasons why a definition of the objectives of the organization must be considered:

1 if goals are unknown it cannot be determined whether or not sufficient information is available to enable planning to take place;
2 should it be desired to determine a correlation between the amount of management information available to the proprietor and the 'success' of the business there must be a measure of 'success'. This would presumably be the ability to achieve goals.

Current thinking on business goals is of two main strands. The first moves from the economic theory concept that businesses attempt to maximize owner's wealth and suggests that due to bounded rationality (Simon 1959) and general inertia, businesses 'satisfice', that is, search until they find a plan which provides satisfactory profits, and operate at that level, rather than continuing to search for greater profits. The second regards businesses as organizations and draws from organization theory to suggest that it is impossible to distinguish the objectives of an organization from the objectives of the participants (Simon 1964, Cyert and March 1969). These individuals bargain, using the resources they bring to the organization, to attempt to acquire their share of personal satisfaction, be it money, power, interesting lifestyle and so on.

Both these ideas seem relevant to small businesses and particularly since, in the very small business, the proprietor does not have to bargain with other parties within the organization it could be deduced that the objectives of the owner are those of the business. If this is accepted, it is necessary to discover whether the owner has enough information to plan, both in the short term and the long term.

Controlling

Control within organizations is a process of monitoring activities to ensure that they are fulfilling organizational goals. It involves scanning the environment for means of fulfilling goals, choosing actions and subsequently checking whether or not those actions have succeeded. In large organizations this also involves defining, co-ordinating and measuring the activities of the various members of the organization. In very small businesses it will focus more on the resources that the proprietor has, to measure the effectiveness of his decisions on the day to day running and long-term decision making of his business.

Organizing

Organizing in large organizations usually relates to deciding who within the organization will perform which task and what resources they need to do it. In particular, for decision makers, it involves identifying which information can be supplied by the internal information system. In a small business with only one decision maker it will focus more on ensuring that the information is to hand (see following section on 'information economics').

Communicating and motivating

Providing regular and complete information to different members of an organization to ensure that they are able to carry out their job effectively, individually and as a team is one of the most difficult tasks of management. Motivating participants to work for the benefit of the organization is equally hard. To effectively respond to participants' communicational and motivational needs requires constant awareness of changes in the organization and in the roles and attitudes of all participants. These issues are less problematic in micro businesses as there is either only one or few participants, so that activities and issues can be discussed regularly and a team spirit can be engendered. However, self-motivation may be a problem for proprietors and a proprietor who has been working alone for some time may find it difficult to work with others.

Information theory: information economics

These approaches regard information as a commodity (e.g. Demski 1980) with associated costs of acquisition and also assume that it is tradeable. The small business proprietor is perhaps in a better position than most to identify the cost of information. He may, for example, decide to spend the evening getting his records up to date rather than going to the pub. The choice of doing the less attractive option must mean that the perceived value of having the information is greater than an evening at the pub foregone. Similar decisions must be made regularly about whether to use time to 'make money' or to seek out ways to improve the business. In a very real sense this is capital expenditure of time. For example, a businessman may be unsure whether or not he can raise prices without loss of business. To find out may involve a lot of time. This time may turn out to have been wasted if he discovers that he could not raise prices, but could have been spent productively if he discovers that he can charge more and ascertains what higher price would be acceptable.

Formal and informal information

Formal information is that which is regularly gathered and disseminated within the organization to help with management processes. Informal information is all the other knowledge acquired by personnel which enables them to do their job effectively but of which there is no written evidence. In a large organization the formal information will be very extensive in order that the different decision makers can be kept in the picture. In a very small business with one decision maker, there is no need to keep others informed, so extensive record keeping may be a waste of resources.

In order to determine the extent of the information available to the small business proprietor it is necessary not only to document business records but also to determine what the businessman knows, irrespective of whether or not it is formally recorded.

THE RESEARCH

Method

In order to identify the management information available to, and used by proprietors, quota sampling was used to obtain a sample of 200 small business proprietors in the West Midlands who ran firms with under ten employees. Firms were chosen at random from the Yellow Pages for Birmingham and the client register of Redditch Enterprise Agency; the sample consisted of 100 proprietors from each.

An interviewer-assisted questionnaire was used to ask about general business questions, specialized questions about record keeping and decision making and the social characteristics of respondents in order to:

1 document the actual management information used by very small businesses and determine the extent to which this may vary by type and age of business and the social characteristics of the businessman;

2 determine the extent to which small businessmen are unable to
obtain necessary information.

Interviewing took place during the latter half of 1989 and early 1990
at a time when the business climate was stable.

Results

Focus of analysis

The responses collected during interviewing have been analysed to
focus on the information available to firms in relation to:

1 transactions with the outside world
2 day to day monitoring of the firm's performance
3 detailed record keeping
4 production, pricing and marketing decisions
5 capital expenditure decisions
6 drawings decisions.

Each of these aspects will be examined in terms of

1 theoretical considerations
2 general survey findings about the practice of firms in the survey
3 the differences between firms.

Sample characteristics

In order to provide an insight into how firms operate and to compare
the practices of different types of businesses, the businesses in the
sample have been analysed in terms of:

1 *age of business* – younger than 3 years (41.5 per cent)[2] or 3 years
and older (58.5 per cent)
2 *size of business* – no employees (31.5 per cent) or at least one
employee (68.5 per cent)
3 *number of owners* – only one owner (61.5 per cent) or more than
one owner (38.5 per cent)
4 *business sector* – manufacturing (37.5 per cent), service (52.0 per
cent) retail (10.5 per cent) (Appendix 2 gives examples of the types
of firm in each catagory)
5 *respondent's perception of achievement* – doing well and very well
(46.5 per cent) or doing adequately and not well (52.0 per cent).

These variables were themselves examined to see if there was any
correlation.

Table 6.1 Analysis of each feature of the business by other features (chi-sq significance test)

	Age of business p	Size p	Number of owners p	Business sector p	Achievement p
Age of business	–	$\leqslant 0.01$	$\leqslant 0.01$	–	–
Size	$\leqslant 0.01$	–	$\leqslant 0.01$	–	$\leqslant 0.05$
Number of owners	$\leqslant 0.01$	$\leqslant 0.01$	–	–	$\leqslant 0.01$
Business sector	–	–	–	–	–
Achievement	–	$\leqslant 0.05$	$\leqslant 0.01$	–	–

Note: p = probability point of the X^2 (Chi-square) distribution.

The analysis indicated that younger businesses were more likely than older businesses to have only one owner and more likely to have no employees. Irrespective of the age of business, where firms had only one proprietor they were more likely to have no employees. Both firms with no employees and firms with only one proprietor were more likely to say that their business was doing adequately or not well. With regard to the business sector, there was no significant difference attributable to age, size, number of owners or achievement.

Transactions with the outside world

Theoretical aspects

Anyone who enters into legally binding contracts with other parties would be advised to keep some documentation of the transaction in order to be able to obtain legal redress should there be any problem or dissatisfaction. Once a person sets up in business the need to keep records of such contracts is even more necessary.

There are legal requirements on firms to keep records of their activities. Sole traders and partnerships, for example, if they become bankrupt, may be charged under the Insolvency Act 1986 with 'failing to keep proper books of account' and all limited companies are required to keep accounting records (s.221 Companies Act 1985). Records are also required for any organization with a turnover exceeding £9,600 in any three month period in order to account for VAT. All trading concerns are required to submit accounts each year to the Inland Revenue to be used as the basis for the assessment of income tax or corporation tax.

Once a firm takes on employees its obligation to keep records becomes even greater as it is necessary to record and calculate

National Insurance and Pay As You Earn deductions. These deductions must be remitted to the appropriate authorities and various forms must be completed at the end of each fiscal year. Furthermore the obligation to operate effectively as a business becomes greater as the livelihoods of employees are also affected.

Apart from the legal requirements, there is a practical necessity to keep sufficient records to ensure that business activity can continue. If a businessman has no knowledge of which bills are outstanding for payment and the extent of his credit and keeps no record of how much is due to him from customers, he will not stay long in business. If he regularly runs into overdraft because he has no record of his bank balance he will either alienate the bank manager or the creditors whose cheques are bounced, or both.

It would follow from this that *all* firms could be expected to keep a record of daily receipts and payments, usually recorded in a 'cash book', and information about those to whom they owe money (creditors) and those who owe money to them (debtors). Some businesses, especially retailers and some service firms, expect customers to pay immediately. They would thus not have debtors and need not keep the relevant records. It is possible that some businesses would not have creditors as they may pay for everything they purchase immediately, particularly if the business is run from home and therefore even electricity and telephone bills will not necessarily be regarded as creditors of the business. However, it would be expected that the large majority of firms would buy some items on credit and therefore have creditors. In order to be able to use records to provide information, for example to monitor cash-flow or send out accounts to debtors, the records need to be written up frequently.

Survey findings

The questions in the survey which related to these aspects were:

1 a general question about which aspects of running a business respondents found difficult. One answer they could choose was 'accounts';
2 the person who writes up the business records, the form in which they are kept and how often they are up-dated;
3 which of the following records were kept:
    (i)   record of cash and bank transactions
    (ii) creditor records
    (iii) debtor records;

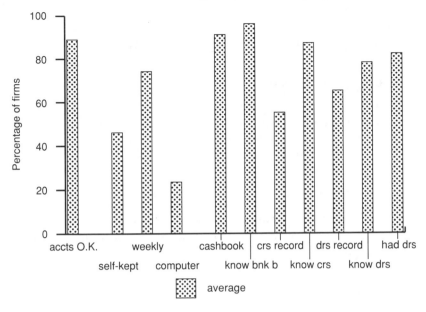

Figure 6.1 Basic record-keeping

4 which of the following information is known:
  (i)   the bank balance
  (ii) the credit period allowed by your suppliers
  (iii) how long your debtors take to pay.

Results (See Figure 6.1)

Accounts

It was found that 88.5 per cent of the survey population did not perceive 'accounts' to be difficult, that is, only 11.5 per cent found them a problem, whereas larger numbers found difficulty with pricing (28 per cent), advertising (22.5 per cent), and market research (14 per cent).

Keeping the books

Does the proprietor keep the books?

When designing the research it was thought that those who wrote up the books themselves might be more in touch with the business.

However, it was discovered that people who knew most about their business did not necessarily write up their own books and only 46 per cent of proprietors had no help with this activity.

How often were the books written up?

Nearly half the firms surveyed (48.8 per cent) wrote up their books daily and another 25.0 per cent wrote them up weekly, which would seem often enough for most types of firm to monitor their affairs. Just over a quarter (26.2 per cent) wrote them up less often.

Is a computer used to keep business records?

Only 23.5 per cent of firms said they used a computer. It might be expected that firms with more complex businesses would be more likely to have a computer.

Cash book (records of receipts, payments)

As expected, the large majority, 91.0 per cent of businesses kept a cash book. Since a cash book would seem necessary for any business to keep a track of the bank-balance and know the amounts which are being received and paid it is of some concern that eighteen respondents did not keep one. Nearly all respondents (96.5 per cent) knew their bank balance. (A profile of the eighteen firms not using a cash book and the seven who did not know their bank balance are listed in Appendix 1 (see p. 228)).

Creditors and debtors

The responses to questions on creditors and debtors show that more respondents had information about their creditors and debtors than had formal records of them. It might be expected that 100 per cent of respondents would *know* details of their creditors and debtors; however, some firms operated on a cash basis and so did not have creditors or debtors. The survey revealed that 89.5 per cent of firms had creditors and the credit payment period was known by 87.5 per cent of respondents, therefore only 2 per cent did not know their credit payment period. However, only 55 per cent of respondents kept creditor records. These records are not essential as creditors will always remind the firm of outstanding debts.

It is much more important to keep debtor records, as the firm will be responsible for reminding their debtors to pay. The survey showed that 81.5 per cent of respondents had debtors however, only 78 per cent knew how long their debtors took to pay. This means that 3.5 per cent did not know how long their debtors take to pay. More significantly only 65.5 per cent said that they kept debtor records, so 16 per cent of respondents with debtors did not have records of them.

Analysis by features of the business (See Tables 6.2 and 6.3)

Age of business

There was no significant difference with regard to basic record keeping between firms which were less than 3 years old and those which were 3 years or older. Interestingly although only 23.5 per cent of the total sample used a computer, they were just as likely to be young firms as older ones. Perhaps this indicates that people just starting up in business are likely to think of using a computer from

Table 6.2 Basic record-keeping: analysis by features of the businesses (chi-sq significance test)

	Young v. older p	No. employees v. some employees p	One owner v. more than one owner p	Business sector p	Doing well v. not doing well p
Accounts not difficult	–	–	≤ 0.01	–	≤ 0.04
Book kept by proprietor	–	≤ 0.01	≤ 0.01	≤ 0.01	–
Written up at least weekly	–	≤ 0.05	≤ 0.01	–	≤ 0.01
Records kept on computer	–	≤ 0.01	–	–	–
Kept a cash book	–	–	–	–	–
Knew bank balance	–	–	–	–	–
Kept creditor records	–	≤ 0.01	≤ 0.01	≤ 0.01	≤ 0.05
Kept creditor payment period	–	≤ 0.01	≤ 0.01	≤ 0.05	≤ 0.01
Kept debtor records	–	≤ 0.05	≤ 0.01	≤ 0.01	≤ 0.01
Knew debtor collection period	–	≤ 0.01	≤ 0.05	≤ 0.01	≤ 0.01
Had debtors	–	–	–	≤ 0.01	≤ 0.01

Note: as Table 6.1.

Table 6.3 Basic record-keeping (% of firms)

	Average	Young businesses[1]	Older businesses	No. employees	Some employees	One owner	More than one owner	Manuf.	Service	Retail	Doing well	Not doing well
Accts O.K	88.5	86.7	89.7	88.9	88.3	83.7	96.1	86.7	89.1	90.5	93.5	83.6
Self-kept	46	48.2	44.4	65.1	37.2	51.2	37.7	56	35.6	61.9	39.8	51
Weekly	73.8	71.1	75.2	63.4	78.1	65.9	85.7	72	70.3	90.2	83.9	64.4
Computer	23.5	24.1	23.1	12.6	28.4	20.3	28.6	26.7	21.8	19	27.9	19.2
Cash book	91	86.7	94	90.4	91.2	91	90.9	93	87.5	100	90.3	91.3
Know bank balance	96.5	96.4	96.6	96.8	96.4	94.3	100	98.7	96.2	90.5	99	94.2
Crs record	55	50.6	58.1	39.7	62	46.3	68.8	69.3	46.5	42.9	62.4	48.1
Know crs	87.5	86.7	88	77.8	92	81.3	97.4	94.7	85.1	76.2	94.6	80.8
Drs record	65.5	61.4	68.4	54	70.8	56.1	80.5	73.3	65.3	38.1	78.5	54.8
Know drs	78	72.3	82.1	66.7	83.2	73.2	85.7	82.7	82.2	38.1	87.1	65.4
Had drs	81.5	79.5	82.9	74.6	84.7	78	87	86.7	86.1	38.1	90.3	74

Note: 1 less than 3 years.

the start, whereas those already established would have to change their routines to begin to use a computer.

Small and larger businesses

The analysis by whether or not firms had employees, shows that businesses with no employees vary considerably from those with employees in nearly every respect. Firms with no employees were significantly more likely to keep their own books, but less likely to write them up very often or to use a computer. They kept a cash book and knew their bank balance, but were much less likely to keep other records or know information about their creditors or debtors. Interestingly they did not seem to perceive accounts to be a problem, so perhaps their affairs are so simple that they can cope or perhaps the lack of employees means that there is less pressure on these firms to ensure that they are effective. Earlier analysis (p. 189) has shown that firms with no employees are more likely to be doing adequately or not well.

One owner and more than one owner

Where there is more than one owner, more people may need access to information so it is perhaps not surprising that firms with more than one owner are more knowledgable and keep more records than firms with only one owner. Firms with more than one owner are significantly more likely to be comfortable with accounts, to write up their books at least weekly, to keep creditor and debtor records and to know their creditor and debtor payment periods. Although not statistically significant, the seven respondents who did not know their bank balance were all firms with only one owner.

Business sector

The analysis shows interesting features about the different types of business. Retailers, who have the least complex decision processes, seemed the most diligent record keepers. They were most likely to keep the books themselves and they all kept a cash book. They were less likely to have creditors and debtors, but 100 per cent of those who had debtors kept a record of them and knew how long their debtors took to pay.

Service firms were most likely to have delegated their book-keeping. Although many of the service firms knew their creditor and debtor payment period they displayed the largest difference between having this knowledge and keeping creditor and debtor records.

Manufacturing firms tend to have the longest lead time between paying for resources and effecting sales, so it would be expected that they would be most likely to need to keep good records. This is borne out by the data which shows that manufacturing firms are most likely to keep creditor and debtor records and to know their payment periods.

Achievement

Firms doing well are more likely to be comfortable with accounts and much more likely to keep their books up to date at least weekly. They are more likely to keep creditor and debtor records and to know their creditor and debtor payment period. Firms not doing well have less debtors than firms doing well. However, a larger proportion of firms doing adequately or not well, who have debtors, fail to keep records of them (87 per cent of firms doing well with debtors kept debtor records compared to 74 per cent of firms not doing well).

Summary

Although this section looks at the most basic record-keeping which it would be expected that most firms would practice, differences emerge from the analyses by the characteristic of the firms and their owners. The firms which were most likely to find accounts difficult were sole proprietors and those doing adequately or not well. The firms in which the proprietors wrote up their own books were likely to be the smallest firms or retailers. The smallest firms, as measured by both employees and owners and those doing adequately or not well wrote up their books less frequently and those with no employees were least likely to have a computer. The most revealing insight from these analyses is that smaller, service firms and those doing adequately or not well do not seem to have adequate debtor records. This feature is worst amongst those doing adequately or not well, where 26 per cent of those with debtors do not keep debtor records. This could be a significant factor in their lack of success and points to a need for small business advisers to ensure that debtor records and collection procedures are in place.

Day to day monitoring of the business

Theoretical aspects

Running a business effectively requires an awareness of how the firm is doing on a day to day basis in order to be able to change strategy or anticipate necessary actions to keep the firm on course. For example, a drop in sales or gross profits or an increase in costs would require the firm to seek extra sales or cut costs to stay in business, or an anticipated sales increase, due to advertising, would require an investment in stock which would itself put a strain on the bank balance, thus requiring the firm to find temporary short-term finance.

The recommended way to plan and monitor a firm's activities is through the use of a budget, drawn up in sufficient detail to enable problems and bottlenecks to be foreseen. The budget is then compared on a month by month basis with the actual activities of the business. Deviations between budget and actual will enable the businessman to change his actions to get back on course. It was felt that, with regard to the micro businesses which were being studied, many would not necessarily prepare budgets but would tend to use previous experience as an *ad hoc* budget to question for example, How did we do last month? What went wrong? What is likely to be different this month? In order to do this proprietors would need to know how their firm was doing week by week.

Due to the level of involvement of the proprietors in all aspects of decision making in these micro firms, much of the information they need is picked up intuitively and formal records may not be used much on a day to day basis (e.g. the florist who has to throw flowers away and the electrician who does not have enough work to fill the day know that they are not doing well). However, periodic formal analysis would seem to be necessary to review activities and calculate the financial impact of the events of the preceding period.

Survey findings

The questions in the survey which relate to these aspects were:

1 questions relating to the preparation of budgets or targets, their comparison with what actually happens and how frequently this is done
2 questions relating to proprietors' knowledge of business indicators:
    (i)   the £ value of sales each week
    (ii) section of the business generating the most sales each week

  (iii) the cost of each item sold
  (iv) the average gross profit percentage
  (v)   the amount of the costs which must be covered before the
        business can make a profit
3 questions relating to the proprietors' awareness of how well the
  firm is doing:
  (i)   even though they may not have firm figures for some areas of
         the business, did they feel that they knew how well the
         business was doing
  (ii)   how often did they calculate how well they were doing
  (iii) how do they calculate how well they were doing?

Results (See Figures 6.2 and 6.3).

Budgets

It was found that 34 per cent of the firms surveyed produced budgets
and most firms (32.5 per cent) then compared the budgets with their
actual performance. To be useful in monitoring the business it would
be expected that the comparison would be done at least monthly,
but only 45.5 per cent of the firms using budgets performed the
comparison this often.

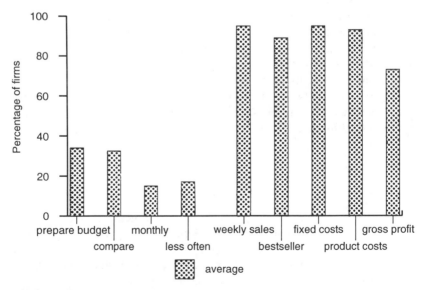

Figure 6.2   Monitoring the business (1)

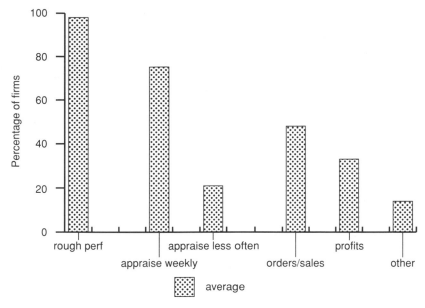

Figure 6.3 Monitoring the business (2)

Knowledge of business indicators

The large majority of firms knew the most important figures needed for monitoring their businesses. The amount of their weekly sales and the amount of fixed costs which must be covered were known by 95 per cent of firms. Product costs were known by 93 per cent and the product or service which was the best seller was known by 89.3 per cent (in some cases this factor was not relevant as there was only one product or service). Only 73 per cent knew their average gross profit percentage (again this may not be regarded as relevant to some, particularly service firms).

Awareness of how the business was doing

Nearly all respondents (98 per cent) said that even if they did not necessarily have firm figures they knew how their business was doing. Most firms (75 per cent) did some calculations to monitor their business weekly, but 21.5 per cent did these monthly, or longer and 3.5 per cent did not reply.

The main methods used by respondents were orders or sales level (46 per cent) or profit calculation (33 per cent). Sales or orders give

Table 6.4 Monitoring the business: analysis by features of the businesses (chi-sq significance test)

	Young v. older p	No. employees v. some employees p	One owner v. more than one owner p	Business sector p	Doing well v. not doing well p
Prepare budgets	–	≤0.05	≤0.01	–	≤0.01
Compare actual with budgets	–	≤0.05	≤0.01	–	≤0.01
Use budgets monthly	–	–	≤0.01	–	≤0.01
Use budgets less often	–	–	–	–	–
Know weekly sales total	–	–	–	–	–
Know best selling product	–	–	–	–	–
Know level of fixed assets	–	–	–	–	–
Know product assets	–	–	–	–	–
Know gross profit	≤0.05	≤0.05	≤0.01	≤0.05	–
Daily aware of rough profits	–	≤0.01	–	–	–
Appraise performance weekly	–	–	–	–	–
Appraise performance less often than weekly	–	–	–	–	–
Appraise by orders/sales	≤0.01	–	–	–	–
Appraise by profit calculation	≤0.01	–	–	–	–
Other	≤0.05	–	–	–	–

Note: As Table 6.1.

an indication of activity levels and thus potential profits, but may hide problems of margins which have been reduced in an attempt to generate sales or increase fixed costs which must be covered by increased sales. Profit calculation is thus the preferred method but clearly requires both more time and expertise.

Table 6.5 Monitoring the business (% of firms)

	Average	Young businesses[1]	Older businesses	No. employees	Some employees	One owner	More than one owner	Manuf.	Service	Retail	Doing well	Not doing well
Prepare												
budgets	34.0	36.1	32.5	22.2	39.4	22.8	51.9	36.0	30.7	42.9	44.4	25.0
Compare	32.5	33.7	31.6	20.6	38.0	21.1	50.6	34.7	29.8	38.1	41.9	24.0
Monthly	15.5	13.3	17.1	9.5	18.2	8.1	27.3	13.3	14.4	19.4	22.6	9.6
Less often	17.0	20.4	14.5	11.1	19.8	13.0	23.3	21.4	15.4	18.7	19.4	14.4
Weekly												
sales	95.0	94.0	95.7	92.1	96.4	94.3	96.1	92.0	96.1	100.0	96.8	93.3
Bestseller	89.3	84.3	91.5	84.1	90.5	86.2	92.2	89.3	85.5	100.0	90.3	87.5
Fixed costs	95.0	96.4	94.0	93.7	95.6	93.5	97.4	93.3	95.1	100.0	95.7	94.2
Product												
costs	93.0	92.8	93.2	92.1	93.4	91.1	96.1	94.7	90.4	100.0	93.5	92.3
Gross profit	73.0	62.7	80.3	61.9	78.1	61.0	92.2	77.3	66.3	90.5	78.5	67.3
Rough												
perf. daily	98.0	98.8	97.4	93.6	100.0	97.6	98.7	98.7	98.1	95.2	98.9	98.1
Appraise												
weekly	75.0	73.5	76.1	66.7	78.9	74.0	76.6	80.0	69.3	85.7	75.3	75.0
Appraise												
less often	21.5	25.3	19.9	27.0	19.0	22.8	19.5	18.7	25.7	9.5	21.5	21.1
Orders/sales	48.5	44.6	51.3	53.1	50.4	46.3	53.2	56.0	47.1	33.3	50.5	48.1
Profits	33.0	43.3	25.6	30.1	34.3	33.3	32.4	26.6	36.5	38.1	29.0	33.7
Other	14.0	14.4	17.1	18.5	10.8	15.9	9.9	12.1	11.6	28.6	11.0	15.4

Note: 1 less than 3 years.

Analysis by features of the businesses (See Tables 6.4 and 6.5)

Age of the business

The only significant differences between young and older businesses were that those running younger businesses were less likely to know their average gross profit and much more likely to use a profit calculation to regularly monitor the firm's performance than a sales or orders figure. These two factors would seem logically linked, as, if the firms do not know their gross profit margins, they would need more accurate monitoring methods to learn from experience their firm's profitability trends. It is also interesting to note that although only 34 per cent of the respondents used budgets, they were as likely to be young businesses as older ones.

Small and larger businesses

Businesses with no employees were less likely to use budgets, only half as many of them used budgets as large businesses. Proprietors of the smaller businesses were less likely to know their average gross profit margins. It was also significant that the four respondents who considered that they did not know roughly how they were doing were all smaller businesses.

One owner and more than one owner

The differences between firms with one or more than one owner are very similar to those observed between small and larger firms but the divergence is slightly greater particularly with regard to the use of budgets and those with one owner were less likely to compare the budgets frequently.

Business sector

The only significant difference between the sectors was the knowledge of gross profit margins. Retailers were the most knowledgeable, then manufacturers and least knowledgeable were service firms. This is not surprising as many service firms do not have a gross margin!

Achievement

The only significant difference is that firms doing adequately or not well are less likely to use budgets and those that do use them are less likely to compare them regularly.

Summary

This section focused on the information needed to run the business and showed that many more proprietors have knowledge of the business than use formalized records. Nearly all have a rough idea of how they are doing and 75 per cent weekly appraise the performance of their business.

The significant differences between firms were in the use of budgets, knowledge of gross profit margins and appraisal methods used by the firms. The smaller firms and those doing adequately or not well were less likely to use budgets. The younger and smaller firms were less likely to know their gross profit margins and younger firms were more likely to use weekly profit calculations rather than relying on sales or orders to monitor their business. Knowledge of gross profit margin also varied. The predominant trends were first, that young businesses and those doing adequately or not well use more detailed techniques to try to monitor their businesses than some firms which are older or doing better; second, that the smaller businesses are less knowledgeable than the larger ones in all areas and some of those with no employees are exercising their entrepreneurial freedom not to know how they are doing; and third, that retailers, who have simpler business processes are more likely to know what is going on than manufacturers or service firms. Advice specifically tailored to helping manufacturers and service firms to calculate weekly profits may be appropriate.

Detailed record keeping

Theoretical aspects

Although, as discussed in the section on transactions with the outside world (p. 189), it is possible for a firm to *manage* with very few formal records, it is likely that established firms will keep more extensive records. To monitor and control costs they are likely to keep details of the level of expenses of the firm. These will be recorded in either an analysed cash book or a nominal ledger or both. In firms which supply goods or services at a future date, there needs to be a record of orders received. Manufacturing and service firms might also need production schedules to organize their work-flow. Stock records may be kept by all types of firms to help with ordering replacement stock and, particularly in manufacturing firms, to monitor the costs tied up in work in progress and finished goods stock.

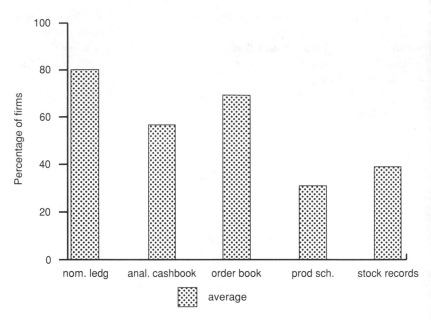

Figure 6.4   Detailed record-keeping

Survey findings

The questions in the survey which related to these aspects were:

Do you keep 1 a nominal ledger
              2 an analysed cash book
              3 an order-book
              4 a production schedule
              5 stock records

Results (See Figure 6.4)

On the whole the firms in the survey were less likely to keep these more detailed records than the basic records, although a nominal ledger was kept by 80.5 per cent which was more than kept creditor records. However, only 56.5 per cent kept an analysed cash book, 69 per cent kept order books, 31.5 per cent kept production schedules and 39 per cent kept stock records.

Table 6.6 Detailed record-keeping analysis by features of the business (chi-sq significance test)

	Young v. older p	No. employees v. some employees p	One owner v. more than one owner p	Business sector p	Doing well v. not doing well p
Nominal ledger	–	≤0.01	–	–	–
Analysed cash book	–	–	–	–	–
Order book	≤0.05	≤0.05	–	≤0.01	–
Production schedule	–	–	≤0.01	≤0.01	–
Stock records	–	≤0.05	≤0.05	≤0.05	–

Note: As Table 6.1.

Analysis by features of the business

Age of business

The only variation from the average in this category was that younger firms were less likely to keep an order book.

Small and larger businesses

Firms with no employees were significantly less likely to keep a nominal ledger, order book or stock records than firms with employees.

One owner and more than one owner

Firms with one owner were significantly less likely to keep a production schedule or stock records than firms with more than one owner.

Business sector

The analysis by business sector is predictable but nevertheless interesting. Manufacturing firms are likely to keep all these types of record and are significantly (over 20 per cent) more likely to keep an order book, production schedule and stock records. Retail firms are the most likely to keep an order book. They do not have production

Table 6.7 Detailed record-keeping (% of firms)

	Average	Young businesses[1]	Older businesses	No. employees	Some employees	One owner	More than one owner	Manuf.	Service	Retail	Doing well	Not doing well
Nom. ledger	80.5	80.7	80.3	69.8	85.4	78	84.4	84	77.2	85.7	86	76
Anal. cashbook	56.5	55.4	57.3	47.6	60.6	52	63.6	66.7	49	57.1	68.8	53.8
Order book	69	60.2	75.2	58.7	73.7	66.7	72.7	85.3	57.7	66.7	71	68.3
Prod. sch.	31.5	30.1	32.5	23.8	35	22	46.8	42.7	28.8	4.8	36.6	26.9
Stock records	39	42.2	36.8	28.6	43.8	33.3	48.1	48	30.8	47.6	45.2	34.6

Note: 1 less than 3 years.

schedules but, along with manufacturers, are likely to keep stock records. Service firms are least likely to keep any of these records except a production schedule, which presumably some maintain to work out priorities in completing jobs.

Achievement

There is no significant difference between firms doing well and firms doing adequately or not well with regard to each individual record. However overall firms doing adequately or not well keep less of these detailed records than firms doing well.

Summary

The numbers of firms keeping order books, production schedules and stock records is lower than for basic records, indicating that many micro businesses do not perceive a need to write down what they plan to do or the stock they have. This lower response rate means that differences are more likely to be observed in the analyses by characteristics.

The most extreme differences occur between small and larger businesses and firms with one owner or more than one, where the smaller firms are less likely to keep these records and firms with one owner are much less likely to use a production schedule. Extremes are also noticeable in the business sector analysis which reflects the nature of the different types of firm. Those firms doing well also show above average use of more detailed records compared to those doing adequately or not well.

Since these more detailed records are more likely to be kept by larger firms and manufacturing firms it can be inferred they are adopted as the activities of the firm become sufficiently complex to warrant the use of these records.

Production, pricing and marketing decisions

Theoretical aspects

Making profits requires that the capacity of the firm and staff time are used as effectively as possible and that prices are set such that customers are attracted to the firm. However, prices should only be as low as necessary to generate business and efforts should be made, where possible, to promote the products or services which generate the most profit for a given input. In order to be able to make decisions on which products to produce and promote the proprietor must have information on:

1 the market for the product and approximate selling price
2 the cost to the firm of the product or service
3 how much of the firm's capacity the product or service uses
4 the fixed costs of the firm which must be met.

From this information it is possible to determine which products the market will buy and at what price, how much is earned by each product, whether some products are more profitable than others in terms of the amount of time used to produce or provide them compared with how much they earned.

Survey findings

The questions in the survey which focus on these aspects were:

1 Did the firm have enough customers?
2 To attract more customers would the firm be prepared to
   (i)   lower price?
   (ii) use more advertising?
   (iii) approach people directly?
3 How are the products priced
   (i)   full cost + per cent?
   (ii) hourly rate?
   (iii) purchase price + per cent?
   (iv) market price?
4 (i)   which of the firm's products sells best?
   (ii) which product is most profitable?

The questions in 4 were used to determine whether the firm was able to exploit its most profitable product.

Question 3 on how products were priced was an open question in which respondents described, in their own words, the way they worked out costs and prices. It would be expected that manufacturing firms would be most likely to use full cost + per cent which involves tracing materials and labour hours used to make a product, calculating their costs and either adding on a proportion of fixed costs and then a percentage for profit, or adding on a larger percentage to cover a contribution to fixed costs and profit. For example a manufacturer of springs and pressings said 'Standard exercise. Look to see how complicated the item is and price accordingly. I consider materials –

copper is expensive, steel will be cheaper. Labour is a set figure, so it is materials + tool variables + overheads + labour plus a percentage.'

The hourly rate (possibly with a charge for any materials provided) is most popular with service firms and involves setting a rate per hour which, assuming a normally busy week will cover the wages of the person doing the work and provide enough to pay fixed costs and earn a profit (this method is commonly used by garages for servicing and motor repairs and by accountants and solicitors, to name a few).

The purchase price + per cent tends to be used by retailers where there is no recommended retail price.

Market price is either the price that competitors are charging, the price the proprietor thinks will be accepted or the price which reflects quality.

The methods described earlier all build up a price from cost. However, if this cost-based price comes to more than the market price and customers are in a position to choose, the firm will not be able to charge the cost-based price. A decision must then be made on whether to adopt the lower, market price or provide a different product or service. Conversely if the market price is significantly higher than the cost-based price the firm can increase price up to the market price. Many firms will thus be in a position of using both cost-based and market price to make their pricing decisions.

Results (See Figure 6.5)

Market knowledge

Only 27 per cent of the firms had enough customers, so the majority (73 per cent) were in a position of needing to attract more. However, only 20 per cent were prepared to lower price to attract more business, and many said they could not afford to do so. In a related question 56 per cent of respondents said they would never reduce prices. More advertising was considered possible by 36.5 per cent and approaching people directly by 62.5 per cent (respondents could have chosen any, all or none of the options).

Analysis by features of the business (See Tables 6.8 and 6.9)

Young and older businesses

Younger businesses were more prepared to advertise than older businesses.

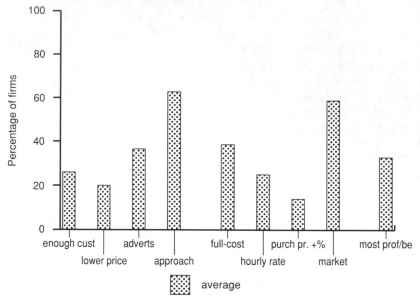

Figure 6.5   Market knowledge and pricing methods

Smaller and larger businesses
No differences.

One owner and more than one owner
No differences.

Business sector
No differences.

Achievement

The firms not doing well were much less likely to have enough customers and were much more prepared to approach people directly than the other firms.

Summary

Even though only 27 per cent of firms had enough customers there was a reluctance to squeeze margins by lowering price or paying for

Table 6.8 Market knowledge and pricing methods (chi-sq significance test)

	Young v. older p	No. employees v. some employees p	One owner v. more than one owner p	Business sector p	Doing well v. not doing well p
Had enough customers	–	–	–	–	≤0.01
Get more customers by lower price	–	–	–	–	–
Advertising	≤0.05	–	–	–	–
Approaching people directly	–	–	–	–	≤0.05
Pricing method					
Full-cost	–	≤0.01	≤0.01	≤0.01	–
Hourly rate	–	≤0.05	≤0.001	≤0.05	–
Purchase price + %	–	–	–	≤0.01	–
Market price	–	–	–	–	–
Most profitable product is best seller	–	–	–	–	–

Note: As Table 6.1.

advertising. Perhaps logically, proprietors who were not busy enough would rather spend their time approaching people directly than cutting their profits by using either of the other methods. The younger firms were the most keen to advertise and those not doing well were even more keen to approach people directly. Getting enough customers is clearly the most difficult problem for micro businesses even in times of relative economic stability. It is very difficult to know what can be done to assist in this area other than discouraging entrepreneurs who propose to enter already crowded markets.

Results

Pricing methods

This open question showed that most firms had a good understanding of the need to build up product cost, but also that they must take account of the market. The responses were:

Table 6.9 Market knowledge and pricing methods (% of firms)

	Average	Young businesses[1]	Older businesses	No employees	Some employees	One owner	More than one owner	Manuf.	Service	Retail	Doing well	Not doing well
Market knowledge												
Enough cust.	27	21.7	30.8	27	27	27.6	26	30.7	26.9	14.3	37.6	17.3
Lower price	20	15.7	23.1	17.5	21.2	18.7	22.1	22.7	18.3	19	19.4	21.2
Adverts	36.5	44.6	30.8	38.1	35.8	36.6	36.4	37.3	35.6	38.1	32.2	40.4
Approach	62.5	69.9	57.3	57.1	65	60.2	66.2	62.7	62.5	61.9	55.9	69.2
Pricing method												
Full cost	38.5	34.9	41	25.4	44.5	31.7	49.3	60	28.8	9.5	39.8	37.4
Hourly rate	24.5	25.3	23.9	34.9	19.7	32.5	11.7	26.7	27.9	0	20.4	26.9
Purch. pr + %	14	18	11.1	14.3	13.9	13.8	16.9	1.3	13.5	66.7	16.1	12.5
Market	59	61.4	57.3	58.7	59.1	62.6	53.2	61.3	57.6	57.1	54.8	62.5
Most prof/ best seller	33.5	32.5	34.2	34.9	32.8	34.1	32.4	26.7	39.4	28.6	39.8	27.9

Note: 1 less than 3 years.

full-cost	38.5%
hourly rate	24.5%
purchase price + per cent	14.0%
market	59.0%
total	136.0%

Where more than one method was mentioned this was in all cases a cost based and a market-based response. This means that 36 per cent of firms used both methods and 23 per cent used only the market price. Some firms described how they used the cost based method as a bench-mark, but then used their judgement with each customer as to how much more (or less) they could charge. The analysis also shows that in only 33.5 per cent of firms was the most profitable product also the bestseller.

Several respondents pointed out that being the cheapest was not always the ideal strategy as price indicated quality in the mind of the customer, so if a high quality product or service was being sold, the price would not only need to reflect that quality but also give the firm the time to ensure that a high quality product was produced. The responses also indicated the job satisfaction enjoyed by the entrepreneur in being able to make the decision on price and resource allocation without being answerable to anyone else. For example, the decision to reduce price to help a relatively poor customer could be made in the knowledge that the lost profit will be suffered by the decision maker alone.

Analysis by features of the businesses (See Tables 6.8 and 6.9)

Young and older businesses

No difference.

Small and larger businesses

Businesses with no employees seemed much more likely to use hourly rate than full cost, whereas firms with employees were the opposite. Hourly rate is simpler to use and is more associated with service firms but the analysis does not show that firms with no employees are more likely to be services.

One owner and more than one owner

Firms with one owner are similarly likely to use hourly rate rather than full cost.

Business sector

As expected this analysis shows the greatest difference between different firms. Manufacturing firms are much more likely to use full-cost than any other method or any other type of firm. Some manufacturing firms use hourly rate. Manufacturers are also most likely to use a market-based as well as a cost-based figure (49 per cent used both methods). Service firms are equally likely to use either full-cost or hourly rate and a few use purchase price + per cent. Retail firms use either purchase price + per cent or market price. The two firms which used full cost were a bridal shop which did alterations and a market retailer selling bread which he made himself, they both also used market price.

Achievement

No difference.

Summary

Product pricing is, for many firms, a difficult activity as the proprietors are constantly unsure of whether the price could be higher or if a lower price would bring in more business.

The analyses show that the major differences were between the different business sectors and largely confirms expectations in those areas. It also shows younger businesses, sole proprietors, manufacturing firms and those not doing well to be seeking the most information on which to base a price.This fits in well with other observations (e.g. information used for monitoring the business) that these firms are somewhat less knowledgeable about their firm's activities, either due to the proprietors' inexperience or the complexity of the firm.

Capital expenditure decisions

Theoretical aspects

Expenditure on items which will benefit the firm for several years and which require a relatively large outlay on the part of the firm mean that two distinct questions must be asked. First, will the firm be more profitable as a result of this purchase and second, how will the item be paid for?

Theoretical texts presume that firms will need to acquire assets at the start of the business which will be part of their 'start up' capital requirement. Thereafter any further capital expenditure decisions will be based on the use of either Discounted Cash-flow (DCF) models or the simpler Payback method. Both methods estimate the future net cash inflows which are expected to result from the capital expenditure. In the DCF model these inflows are then discounted at the expected cost of capital for the number of years into the future that they are expected to be received. If the result of this calculation is more than the initial outlay the expenditure is financially viable. Payback calculates the number of years it will take for the capital expenditure to generate net cash-flows equal to the initial outlay (see Samuels, Wilkes and Brayshaw 1990). An assumption underlying these methods is the objective of maximizing ordinary shareholders' long-term wealth.

Strategic management literature would point out that the calculation of appropriate financial estimates is only part of the process of deciding on investment and that the asset acquired must fit in with the forward strategy which management has identified. It was felt likely that the micro businesses surveyed would not use DCF calculations. However, it was not known how they did operate.

Survey findings

The question asked on this topic was, 'If you wish to purchase new equipment for your business, please describe how you would decide whether it was worthwhile?'.

Results (See Figure 6.6)

The responses showed a very real understanding of the issues involved because in most cases the capital would have to be provided by the proprietors, either by introducing more capital, foregoing drawings or obtaining a loan secured on their personal property. The responses again demonstrate how the owners' involvement in the business makes the decision procedure informal and it is simplified by there being one decision maker, one beneficiary and one financier.

It appeared that firms acquired their 'necessary' assets over time, rather than at the start of their business. Where businesses start at home it is likely that some assets will have been acquired while the business was still a hobby. This is another area where the separation of the business from the private life of the proprietor is more in the

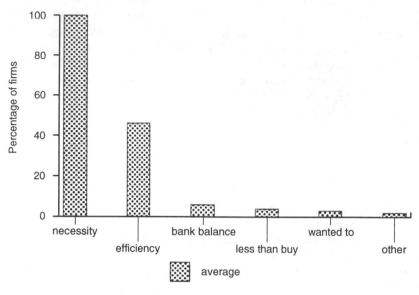

Figure 6.6   Capital expenditure decisions

minds of the business analyst than the proprietor himself. For a large
group (36.5 per cent) 'necessity' was the governing factor. For
example, a vehicle repairer said 'If we could not manage without it
we would buy it, but if we would only use it once or twice we would
think about hiring it' and a gymnasium sports business commented,
'I have had to update the range of gym equipment to keep ahead of
the competition. This is essential to running the business'. These
responses indicate that in some cases well after start up there is no
choice as to whether or not to make the investment.

The highest number of responses (46 per cent) cited greater
efficiency. In some cases the search for efficiency seemed to describe
a subjective awareness, for example, a specialist packaging firm said
'contribution to the efficiency of the business . . . would it move us
a small step further along' and an upholsterer said 'If I could increase
production I would purchase equipment regardless.' Other responses
indicated that some calculations would be done to estimate the level
of increased efficiency, for example, a supplier and fitter of wind-
screens said 'I would take into account the capital outlay and how
long it would take to recoup that' and a health food store proprietor
'I weigh up the resulting profit, either time or cash.' All the responses
in this category indicated a reluctance to spend unless tangible

benefits could be seen, the lack of formal paper analysis appeared to be more a function of not having to justify the expenditure to anyone else than a lack of analytical thought.

The next two categories do not have many respondents, but focus on the problem of finance. The bank balance was regarded as the main criterion for 6 per cent of the respondents although it was also mentioned as a factor by others. For example, a health, safety and security consultant said 'I would not go into debt to purchase equipment.' However, the proprietor of a nursery perceived the bank balance as rationing her desire to buy more to improve her nursery: 'My accountant thinks I spend too much on equipment. I buy two new books each month for variety and other equipment as necessary. The next major purchase (tables) cannot be justified at present'. The personal fulfilment of running a well-equipped nursery clearly outweighed the desire to make money.

Leasing as a financial option was considered by 4 per cent of respondents. A few proprietors (3 per cent) said they would buy new equipment if they wanted to.

Analysis by features of the business (See Tables 6.10 and 6.11)

Young and older businesses

Younger businesses were more likely to regard necessity and an available bank balance as the most important criteria, whereas more of the older firms had obtained their necessary equipment and would buy more if efficiency could be improved. The older firms were also

Table 6.10 Capital expenditure decisions (chi-sq significance test)

	Young v. older p	*No. employees v. some employees* p	*One owner v. more than one owner* p	*Business sector* p	*Doing well v. not doing well* p
Necessity	$\leqslant 0.05$	$\leqslant 0.05$	–	–	–
Efficiency	$\leqslant 0.05$	$\leqslant 0.05$	–	–	–
Bank balance	$\leqslant 0.05$	$\leqslant 0.05$	–	–	–
Lease then buy	$\leqslant 0.05$	$\leqslant 0.05$	–	–	–
Wanted to	$\leqslant 0.05$	$\leqslant 0.05$	–	–	–
Other	$\leqslant 0.05$	$\leqslant 0.05$	–	–	–

Note: As Table 6.1.

Table 6.11 Capital expenditure decisions (% of firms)

	Average	Young businesses[1]	Older businesses	No. employees	Some employees	One owner	More than one owner	Manuf.	Service	Retail	Doing well	Not doing well
Necessity	36.5	41.0	33.3	39.7	35.0	37.4	35.1	41.3	34.6	28.6	37.6	35.6
Efficiency	46.0	39.8	50.4	39.7	48.9	43.0	50.7	48.0	43.2	52.5	48.4	43.3
Bank balance	6.0	10.8	2.6	11.1	3.6	7.3	3.9	2.6	8.6	4.8	3.2	5.8
Lease then buy	4.0	1.2	6.0	0.0	5.8	3.3	5.2	5.3	2.9	4.8	2.2	5.8
Wanted to	3.0	2.4	3.4	4.8	2.2	4.1	1.3	1.3	4.8	0.0	4.3	1.9
Other	2.0	1.2	2.7	0.0	2.8	1.6	2.6	0.0	2.9	4.8	2.1	1.9

Note: 1 less than 3 years.

more likely to consider leasing, perhaps this had not occurred to the proprietors of younger businesses as an alternative.

Small and larger businesses

Firms with no employees are more likely to be driven by necessity and the availability of a bank balance. None of these smallest firms mentioned leasing as a way of obtaining equipment.

One owner and more than one owner

No difference.

Business sector

No difference.

Achievement

No difference.

Summary

The results show that most firms used an efficiency criterion before buying capital items, however in nearly as many cases necessity was the reason for purchase. Necessity was much more likely to be the reason in younger firms and firms with no employees.

Drawings

Drawings are amounts taken out of the business by the proprietors in anticipation of profits. If the firm is an unincorporated business they will be called 'drawings' and if a limited company they will be called 'directors' emoluments'. However, whatever the legal status of these micro businesses, the proprietors are likely to be taking money out of the business on a regular basis for their personal needs. The method of arriving at the decision on how much to take can throw light on the way proprietors make financial decisions about their firms.

Survey findings

The question in the survey relating to this aspect was, 'How do you decide how much to take out of the business for your own use'. This was an open question and it was hoped that respondents would describe how they calculated how much the business could afford. However, although the responses were less detailed than was hoped, they show several different ways of coping with the decision. The responses were able to be categorized into:

1 Salary, which meant the same amount taken each week or month. In some cases the response was 'a salary' or 'a set wage'. Others were more specific and said, for example, 'regular sum taken out determined by the profit during the previous accounting period' or 'Takes a fixed wage regardless of how the business is doing'.
2 Wage plus bonus, which meant that a specific amount was taken out regularly and bonuses were taken if justified by the profits.
3 What the business can afford. This covered any response which indicated that drawings varied week by week depending on business activity, for example, chiropodist 'As this is my own business whatever is left after paying everything is mine'. In some cases it included the decision not to withdraw as much as was earned, for example, nationwide transport operation 'Minimum possible. Looking at five year plan before returns are expected. Spending a lot in time and effort for small returns at present'.

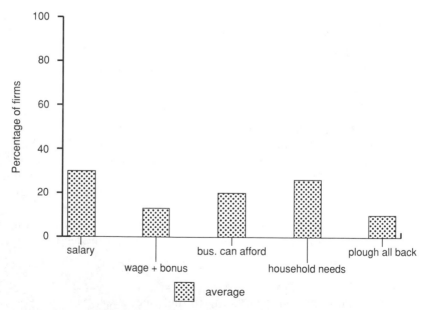

Figure 6.7   Drawings method

4 Household needs. Responses in this category indicated that they only took enough for basic living expenses, in some cases, where the business was not the main means of support this was not necessarily a living wage.
5 Ploughing everything back, which meant that at present all earnings were being put back in the business. The businesses in this category were in all cases not the main means of household support of the proprietor. However, the firms were not necessarily doing badly.

Results (See Figure 6.7)

The largest group 29.5 per cent, took a salary, but all the other respondents used some method to adjust their drawings to the business needs. A wage plus bonus was taken by 13 per cent, what the business could afford was the criterion for 20.5 per cent. The amount needed for household expenses was taken by 26.5 per cent and 9.5 per cent were ploughing everything back.

Analysis by features of the business (See Tables 6.12 and 6.13)

Young and older businesses

Younger businesses are considerably more likely to be taking only enough for their household requirements or ploughing everything back into the business. Older businesses on the whole have achieved a more stable situation enabling more of their owners to take out a salary or wage. However, it is interesting to note that five proprietors of firms over 3 years old are still ploughing back all profits.

Table 6.12 Drawings method (chi-sq significance test)

	Young v. older p	*No. employees v. some employees* p	*One owner v. more than one owner* p	*Business sector* p	*Doing well v. not doing well* p
Salary	≤0.01	≤0.01	≤0.01	–	–
Wage + bonus	≤0.01	≤0.01	≤0.01	–	–
What business can afford	≤0.01	≤0.01	≤0.01	–	–
Household needs	≤0.01	≤0.01	≤0.01	–	–
Plough all back	≤0.01	≤0.01	≤0.01	–	–

Note: As Table 6.1.

Table 6.13 Drawings method (% of firms)

	Average	Young businesses[1]	Older businesses	No. employees	Some employees	One owner	More than one owner	Manuf.	Service	Retail	Doing well	Not doing well
Salary	29.5	21.7	35	12.7	37.2	21.1	42.9	36	26.9	19	34.4	25
Wage + bonus	13	9.6	14.4	14.3	12.4	10.6	16.9	12	14.4	9.5	17.2	9.6
Bus. can afford	20.5	18	22.2	27	17.5	24.4	14.3	21.3	22.1	9.5	19.4	22.1
Household needs	26.5	32.5	22.2	28.5	25.5	31.7	18.2	26.7	25	33.3	19.3	31.7
Plough all back	9.5	16.9	4.3	12.7	8	11.4	6.5	2.7	11.5	23.8	9.6	10.6

Note: 1 less than 3 years.

Small and larger businesses

Firms with no employees are least likely to take a salary and much more likely to use a method based on business success or personal need. This would seem to indicate again that, for these self-employed people, there is little distinction between their business and their private affairs, which is probably the case, particularly for those who work from home.

One owner and more than one owner

Firms with more than one owner show the highest tendency to take a salary or a wage plus bonus. This would be a likely decision process as a practical way of avoiding having to regularly agree on how much each owner should take out of the business. It is likely that those firms with more than one owner which gave other responses were owners who share the same household. Firms with only one owner were much more likely to take drawings either related to the business needs or their own needs. They are also more likely to be ploughing it all back.

Business sector

No difference.

Achievement

No difference.

Summary

Firms which take a salary or wage plus bonus tend to be those which are older or have more than one owner. Thus firms which are established and meeting more complex decision situations are more likely to take a regular amount and simplify one area of their business activity by doing so. Single owners in less complex businesses and firms which are young or not doing well have a more flexible approach related to business and personal needs.

CONCLUSIONS AND DISCUSSION

Summary by characteristics of the firms

Young and older businesses

Young (under 3 years) businesses were found to be more likely than the average (by 21 per cent) to have only one owner and no employees (38 per cent more than the average). However, although subsequent discussion will show that these factors are likely to result in differences, these did not appear in the analysis of younger and older firms, indicating that those small firms which are younger are more likely to keep more books and have more knowledge of their businesses than their counterparts running older businesses of the same size.

The various analyses show little difference between young and older firms. The younger firms were almost as likely to keep necessary books and records, to have a computer and to keep budgets. They were less knowledgeable about the profitability of their products, but were more prepared to be involved in activities which would give them more information such as: 1) monitoring their business by use of a profit calculation, rather than relying on sales levels as a measure; and 2) using both cost-based and market-based information to set price. They were less likely to have enough customers, but were more interested in advertising than older firms to obtain more customers.

The major differences appeared in their capital expenditure and drawings processes. The younger firms' financial constraints meant that they were most likely to buy only necessary equipment and were constrained by the bank balance. This also meant that they were less likely to take a salary and more likely only to take amounts for household needs or to plough everything back into the business.

Small and larger businesses

One owner and more than one owner

These two categories indicate different aspects of size. The small and larger businesses feature the absence or existence of employees. One owner and more than one owner indicates different issues in information dissemination for decision making. However, there was a tendency for those with only one proprietor not to have employees so there is a likelihood that the very smallest firms are those with no employees and the largest those with more than one proprietor. The various analyses show that the greatest differences occur due to the size of firms.

With the exception of keeping a cash book the smallest businesses are less likely to keep records, and those they keep they write up less often. They are less likely to have a computer, prepare budgets or work out how they are doing at least monthly. However, they are just as likely to have enough customers and be as diligent in setting price as larger firms. The smallest businesses are also more likely to only buy necessary equipment and gear their drawings to the business and their personal needs. These tiny firms are also more likely to be doing adequately or not well.

The smallest firms are managing with less records and knowledge than larger firms. This may be contributing to their lack of achievement or it may be that those doing adequately or not well stay small and affect the analysis. If the smaller firms wish to grow they will need to develop more records and methods of information gathering.

Business sector

The firms in the different sectors operate in different ways and therefore have different requirements. These have been analysed and explained in earlier sections. Taken as a whole the analyses do not show any business sector as recording or knowing more than another, but reflect the needs and behaviour of each sector.

Achievement

Achievement was analysed by the subjective responses which respondents gave to the question do you think your firm is doing 1) very well; 2) well; 3) adequately; 4) not well. These groups have been amalgamated into two groups those doing 'well' and those doing 'adequately or not well'. The analysis shows that on the whole firms doing well keep more records, write them up more frequently and know more about the business than those doing adequately or not well.

Firms doing adequately or not well show some worrying features. They are more likely to find accounts difficult. They are less likely to keep formal debtor or creditor records or know their debtor and creditor collection period than any other category, even the smallest firms. They are less likely than firms doing well to write up their books or work out how well they are doing frequently and they are less likely to prepare budgets or keep additional detailed records.

They are the only group with significantly less customers than the average. They seem to be doing their best to develop their market. They would be keen to approach people directly and they tend to use both cost-based and market-based information to set their price, but they cannot exploit their best selling product.

The firms which are doing adequately or not well reveal in almost every analysis lower response rates (even where this is not statistically significant with regard to each observation). The only exceptions were efforts to obtain customers and the drawings method used where they were more likely just to be taking out sufficient for their household needs or ploughing everything back.

The research does not prove that lack of business records and awareness of key business factors is the reason for lack of achievement, but it does show that firms not doing well are less likely to have this knowledge and these records.

Conclusions and areas of focus for advice

This chapter is based on a survey of how accounting record-keeping is used by micro businesses. It describes and analyses the responses by relating them to the management accounting decision processes which firms undertake as part of their business process. It shows that much of the day to day knowledge required by proprietors is greater than that which is written down. However, there are differences between firms in both how knowledgeable they are about their businesses as well as the recorded information. Overall it was discovered that first, where there was more than one proprietor or employees, firms not only kept more records but also knew more about their businesses and second, firms doing well were more likely to keep more records and know more about their businesses than those doing adequately or not well.

In respect of the most basic record keeping it was discovered that there were a few firms which did not keep records and did not have knowledge of how their business was doing (Appendix 1 profiles those who did not keep a cash book or know their bank balance). Some very small firms are able to manage with very few records. However, a worrying discovery is that a number of firms (16 per cent) who had debtors did not keep debtor records. The firms most deficient in this area are smaller, service firms, and those doing adequately or not well. This could be a significant inhibitor to success as the firms may not be collecting all their sales revenue.

Advice is needed on keeping debtor records and chasing up debts.

The study highlighted the fact that although most businessmen were aware of their firm's activities, information tended not to be formalized. Micro businesses tend to use informal information for their day to day decision processes. The existence of only one decision maker means that information does not necessarily need to be written down for access to others. Proprietors were mentally monitoring their achievements on a daily basis and 75 per cent used a formalized weekly appraisal. However, the larger and more successful the firm the more likely it was to have more formalized records. Younger businesses though predominantly small were as likely as older businesses to keep more extensive records, to budget and to calculate in detailed their weekly performance, whereas small firms and those doing adequately or not well were less likely to be so diligent. This could indicate that when firms are young and the proprietors are less experienced in how their decisions impact on profits they need *more* regular information to help future decision making. *Advice is needed at start up to help firms to develop simple information gathering methods quickly. This would seem important in preventing failure and would also mean that systems were already in place should the business grow rapidly.*

Responses indicate that only 34 per cent of firms use budgeting and only 33 per cent regularly calculate profits to monitor their firms performance, whereas 48.5 per cent use the less reliable measure of sales or orders. *A well designed proforma which would help firms to budget and calculate profits regularly could find a market.*

Although firms had knowledge of their business activities which were not recorded, when asked which aspects of their businesses they knew, it was found that the more simple the business process involved, for example, in retail firms, the more likely it was that the proprietor would have full information. Manufacturing firms which have the greatest lead-time between sales and profits were most likely to use a sales figure to measure performance rather than a profit figure. *Advice to manufacturers on weekly profit calculations could help with their business monitoring.*

A smaller number of firms considered it necessary to keep detailed records rather than basic records. These were mostly larger firms and those in the manufacturing sector. It would appear from this that additional records are adopted as firms become more complex.

Market analysis indicated that most firms did not have enough customers, those doing well were much more likely to have enough customers than those doing adequately or not well. The vast majority were not prepared to reduce price in order to obtain more business and some were reluctant to spend money on advertising; younger businesses were the most prepared to do so. More were prepared to approach people directly, particularly those doing adequately or not well. A lack of customers is clearly the most difficult problem for micro businesses and new businesses should be discouraged from entering already crowded markets.

With regards to pricing, all firms have at least one method of establishing a price and some had two. This process appeared to be better understood than anticipated. However, firms were still unsure whether the price they were charging was correct as they were unsure of the effects on demand of a change in price. *It would appear that small firms are most likely to succeed in niche markets where higher prices may be chargeable. Advisers should encourage small firms not to enter crowded markets and to differentiate their product or service to survive. Advice on how to conduct basic market research may also be useful.*

Capital expenditure decision-making showed differences between younger and smaller firms compared to older and larger ones. The smaller and younger firms tended to focus on necessity. The ability to fund the purchase from the resources of the firm or the owners appeared to be an important focus for younger and smaller firms and these issues were raised again when investigating the decision to take drawings. The older and larger firms were more likely to use an efficiency criterion for capital expenditure and to have formalized their drawings to take a regular amount each week. *These predictable findings emphasize again the need for new firms to ensure that they can 'weather' the early years by having sufficient initial capital.*

APPENDIX 1

Who does not keep a cash book

It is interesting to discover how those who did *not* keep a cash book managed their affairs. Of the eighteen proprietors who did not keep a cash book, fourteen used one of two methods of substituting for this, thus keeping some check on their bank balance and checking on receipts and payments.

1 Ten respondents kept a running total in their cheque books. Investigating these respondents is very interesting as it shows that eight of the businesses were under 3 years old and had no employees, five were doing well or very well and three adequately or not well. It is to be hoped that they will all introduce more comprehensive records as they develop. The other two were more than 3 years old and were larger businesses but were only doing adequately or not well. These would seem to be cases for an improvement in record-keeping.

2 Four respondents relied on their bank statement, which is not necessarily a very up-to-the-minute method of determining the bank balance. These four all worked from home with no employees, three were well established service businesses, run by men, which were doing very well or well, so they presumably are managing without this information. The fourth is a woman who has only been going three months making patchwork items and is aware that she needs more records.

There were four more respondents who did not have a cash book. One respondent was doing only adequately, but regarded the business more as a hobby and didn't seem to care; his wife worked in a bank and dealt with all the family business affairs. He also did not know his bank balance.

A second respondent was everyone's idea of the sole proprietor. He is a plumber who had been in business for twenty-six years, he has a big envelope containing batches of receipts which are written up quarterly by his wife for the V.A.T. claim; however, he reckons to know everything that is going on and keeps a continuous watch on the bank balance.

The final two respondents had extensive other record keeping (in fact one was currently transferring his records onto computer); however, they did not say that they had a cash book.

Who does not know their bank balance?

Nearly all the respondents (96.5 per cent) said that they knew their bank balance, it is therefore interesting to look at the seven who did not.

These people were all sole proprietors. Four proprietors said that they did not need this information, although all of them were doing adequately or not well. Two of the businesses were market traders dealing mostly in cash and another ran a secretarial business from home. In each of these cases, the business was not the main means of support and there were no full-time employees. The fourth business is a manufacturer of sails which has been going for ten years and employs four full-time staff. The books are kept on a computer by himself and his wife, who is a chartered accountant. He says that cash flow is the most difficult area of running the business, but he does not need to know his bank balance!

One other said he would like to have this information regularly. He is an interesting case because he has apparently adequate book-keeping methods but does not seem to be able to use them to obtain information. The business is a mobile video hire shop which has been going two years and is the main means of financial support. The business is not doing very well. The proprietor had been advised by the enterprise agency to keep a Simplex cash book (Simplex, if kept correctly is designed to enable the user to calculate a weekly bank balance). This is written up monthly by his accountant. It seems a pity that he was not advised to write up his books more regularly and to understand the information he could obtain from the books.

The sixth respondent who does not know her bank balance is a 60-year-old lady speech therapist who works from home, is doing very well, does not have any employees so presumably knows she has enough money to meet her out-goings without constantly worrying about her bank balance. The seventh is the man who does not keep a cash book either.

APPENDIX 2   EXAMPLES OF TYPES OF BUSINESS

Manufacturing

Upholsterer
Studio Potter
Industrial Packaging
Manufacturers of Aluminium and UPVC Windows
Manufacturers of Enamel Pill Boxes
Chocolate Manufacturing
Neon Sign Manufacturer

Service

Chimney Sweep
Electrical Contractor
Speech Therapist
Day Nursery
Catering
Health and Fitness
Holiday Company Specializing in small group safari type tours

Retail

Second Hand Bookseller
Florist/Greengrocer
Retail Butcher
Retail Carpets
Radio Controlled Model Cars
Health Food Store
Camping and Leisure Equipment Sales – Retail and Hire

NOTES

1 There is an alternative view of business goals, that suggests that actions
  precede goals (March 1972 and Weick 1977). When an action has been
  taken it is evaluated to determine what it was intended to accomplish. This
  might lead to an interesting line of enquiry based on the idea that because
  certain information was/was not readily available the business reacted in
  a particular way.
2 (per cent) indicates per cent of population.

REFERENCES

The Bolton Report (1971) *Small Firms Report of the Committee of Inquiry
  on Small Firms*, HMSO, London.
Carsberg, B.V., Page, M.J., Sindall, A.J. and Waring, I.D. (1985) 'Small
  Company Financial Reporting', Prentice Hall International.

Curran, J. (1986) *Bolton Fifteen Years on: A Review and Analysis of Small Business Research in Britain 1971–1986*, Small Business Research Trust.

Curran, J. (1989) 'The role and impact of small business research' in *The Role and Contribution of Small Business Research*, proceedings of the 9th National Small Firms Policy and Research Conference, 1986, ed. Rosa P. Gower.

Cyert, R.M. and March, J.G. (1969) *A Behaviourial Theory of the Firm*, Prentice Hall International.

Daly, M. (1991) 'The 1980s – A decade of growth in enterprise', *Employment Gazette*, March, pp. 109–34.

Demski, J.L. (1980) *Information Analysis*, Addison Wesley, Reading, MA, 2nd edition.

Drucker, P. (1974) *Management: Tasks Responsibilities and Practices*, Heinemann, London.

Hankinson, A. (1983) *The investment problem: a study of investment behaviour of South Wessex Small Engineering Firms 1979–1982*. SSRC supported project, Dorset Institute of Higher Education Monograph.

Hankinson, A. (1985) 'Output determination: a study of output determination of Dorset–Hampshire small engineering firms 1983–1985', ICMA supported project, Dorset Institute of Higher Educations.

Holmes, S. and Nicholls, D. (1989) 'Modelling the accounting information requirements of small businesses', *Accounting and Business Research*, 19, 74, 143–50.

Lewis, J. and Toon, K. (1986) 'Accounting for growth – small firm growth and the accounting profession', The Small Business Research Trust.

McMahon, R.G.P. and Davies, L.G. (1991) 'Financial reporting and analysis in smaller growth enterprises: evidence on practice in the north-east of England', Accounting and Finance Research Paper 4/91, The Flinders University of South Australia.

March, J.G. (1972) 'Model bias in social action', *Review of Educational Research* 42(4), 413–29.

Nayak, A.M. and Greenfield, S.G. (1991) 'Very small businesses. Accounting information and business decisions – a question of definition', paper presented at the UKEMRA Small Firm '91 Conference.

Robertson, M.R. (1988) 'The great mismatch', Leeds Polytechnic, Department of Management, Leeds.

Samuels, J.M., Wilkes, F.M. and Brayshaw, R. E. (1990) *Management of Company Finance*, Chapman & Hall, London, Fifth Edition.

Simon, H.A. (1959) 'Theories of decision making in economics and behaviourial science', *American Economic Review*, 49(3).

Simon, H.A. (1964) 'On the concept of Organisational Goal', *Administrative Science Quarterly*, June, pp. 1–22.

Stanworth, J. and Gray, C. (eds) (1991) *Bolton 20 Years On: The Small Firm in the 1990s*, Small Business Research Trust, Paul Chapman Publishing, London.

Weick, K.B. (1979) *The Social Psychology of Organising* Addison-Wesley, Wokingham, Bucks, 2nd edition.

Wootton, C. and Templeman, G. (1985) *Small Businesses: Management Information Needs*, CIMA Research Publication.

7 Incorporating the micro business: perceptions and misperceptions

*Judith Freedman and Michael Godwin**

INTRODUCTION

The limited liability company has come to be associated with entrepreneurship within our business culture. This is a notion which is reflected in the language and preoccupations of policy makers. A typical statement comes from the Institute of Directors:

> The limited liability company is the unsung hero of all free enterprise economies. Its structure provides a unique, flexible, efficient means of bringing together capital and expertise, raising finance, limiting risk, stimulating investment and channelling economic activity. In Britain, the competitive market has streamlined time consuming registration procedures so that an entrepreneur can acquire and begin to trade with a private limited company, ready made, in five minutes for a payment of £100.
>
> (IOD 1986)

Thus, not only is the corporate form praised, but also approbation is given for its cheap and easy availability in the UK. Elsewhere in Europe, incorporation is more expensive and formalistic, requiring a payment of minimum capital in most countries. Even so, availability of the corporate form to small firms is seen as essential and this policy led to the adoption of the Twelfth Company Law Directive (EC 1989), requiring Member States to permit the incorporation of single-member companies. The Commission's explanatory memorandum comments on the need to promote 'the access of individual entrepreneurs to the status of company, which represents the best framework for business development in the internal market'.

This enthusiasm for the limited liability company seems to extend, therefore, to incorporation by even the smallest firms. Concern is expressed about the appropriateness of the full blown corporate form for very small firms and whether the benefits of incorporation outweigh the burden for such companies, but this often leads to suggestions for a simplified corporate form for small companies rather than a move from corporate status to unincorporated (Chesterman 1982; DTI 1981; Farrar *et al.* 1991, pp 534–6; Freedman and Godwin 1991). However, there is also a view that incorporation in the form currently available in the UK can impose an administrative burden on small businesses and this has led to some suggestions that unincorporated form may be more efficient and preferable to incorporation for such firms (DTI 1988; DTI/IR 1987: 2).

This chapter compares the reasons given by very small business owners for incorporating, and their level of satisfaction with incorporation, with the advantages of incorporation put forward in legal and economic literature. The authors argue that some of those who are incorporating businesses with so-called limited liability may not be benefiting from this 'privilege', but rather may be victims of a culture which encourages incorporation indiscriminately, and a non-neutral tax system which results in unintended distortions. There is a cost attached to incorporating which may not be outweighed by the actual benefits in some circumstances and it is a cost which is suffered both by the business owner and the wider community.

A further theme of this chapter is that the misunderstandings surrounding the reasons for incorporation and the nature of the incorporated firm lead policy makers (and small business researchers) to focus their attention unduly on incorporated firms to the exclusion of unincorporated firms. This feeds the notion that incorporation is essential for serious business formation (for example, see Robson 1991) and helps to create a pressure to incorporate in conditions where this does not enhance efficiency and is not, ultimately, helpful to those involved. It may also lead to loss to the community if it encourages excessive risk taking. It is argued that it is incorrect to view incorporated firms as a homogeneous group; a wide range of firms incorporate for a variety of reasons and there is limited value in targeting incorporated firms as one group, either in research or policy making. Figures are presented (Statistical Appendix, p. 270) which show that between 85 and 90 per cent of UK companies fall within the Companies Act definition of small companies[1] and it is contended that this category is too broad to be a valuable test of size for many purposes. Further research is required to break down this category for the purposes of applying deregulatory measures and to facilitate future research. Finally the chapter suggests one test, based on the empirical data, which may distinguish between qualitatively

234 *Judith Freedman and Michael Godwin*

different types of company: the importance of share capital at incorporation.

PERSPECTIVES ON INCORPORATION

Small business literature

The question of the legal form of organization of small businesses, that is whether they are sole traders, partnerships or limited liability companies, is not often given attention by small business researchers. However, it does sometimes appear as an issue in connection with investigations into levels of growth amongst small firms. Thus Hakim has found that no-growth firms are typically unincorporated whilst fast growth firms are typically incorporated (Hakim 1989). Batstone has suggested that companies might be targeted for assistance over other business forms since companies are likely to be larger and offer greater benefit in terms of new employment (Batstone 1991). Robson sees unincorporated firms as of interest only as a stepping stone towards attaining corporate status (Robson 1991), whilst Storey suggests that corporate status at start up is linked to a greater likelihood of obtaining bank lending and having more employees (Storey 1992). In work which focuses on the importance of job creation by small businesses as the source of their importance to the community, those *unincorporated* firms which consist of sole traders with no employees ('own account' workers), may be relevant only in so far as they are 'apprentice' employers (Storey and Johnson 1987). It would appear that, in so far as the issue is considered at all, the limited liability company is of more interest to the small business research community than are unincorporated firms; as with the statements quoted above, so here, limited liability companies and entrepreneurship have become equated, or at least associated.

Economic literature

Discussions on the benefits of incorporation in the economic literature tend to focus on larger companies. So King (1977) writes:

> What then is a company, and what is the role of the legal system in regulating its behaviour? Put very simply a company is an association of a number of individuals for the pursuit of economic gain. This immediately differentiates a company from the classical concept of an owner-managed firm
>
> (1977: 24)

Limited liability provides the incentive for individuals to combine their resources to form a joint enterprise on a scale which would be too large for the firm to be based on the essentially personal nature of a partnership . . . Since the risk of investing in such companies is, then, strictly limited, this both encourages individual investors to put their money into risky ventures, and also makes it easier for the company to raise capital to finance its expansion.

(1977: 26)

Since the focus of King's study is the behaviour of public companies, this is a reasonable model within his context. However, difficulties arise when this pervasive view of the corporation, with its emphasis on division of ownership and control, is applied in a small business context, since it ignores the existence of a vast number of owner-managed, very small companies, many of which are in fact single-person companies.[2] King's description does not cover the majority of companies in the UK, but reveals a common perception of the corporate form.

The literature on transaction and agency costs so widely relied upon to illuminate the structure of the firm as a 'nexus of contracts' (Coase 1937; Jensen and Meckling 1976; Fama and Jensen 1983) is explicit in its recognition of the distinctions between different legal forms and the different types of company. Thus, Fama and Jensen distinguish organizations from one another by reference to the characteristics of residual claimants (that is those who have contracted for the rights to net cash flows after fixed payoffs). Open, or large, public, corporations are distinguished from closed corporations that are generally smaller and have residual claims that are largely restricted to internal decision agents. However, it is assumed that, whilst proprietorships will have a single residual claimant, partnerships and closed corporations will have multiple residual claimants. Fama and Jensen state that in proprietorships, partnerships and closed corporations, because of the restriction of residual claims to important decision agents, agency problems that arise because of separation of risk-bearing and decision functions in open corporations are avoided. 'Thus costly mechanisms for separating the management and control of decisions are avoided'. However, they acknowledge that in partnerships and closed corporations, some mechanisms for resolving conflicts among residual claimant decision makers are required.

These descriptions ignore the prevalent 'single-person company' (that is the company with one *beneficial* owner, regardless of whether it has an additional nominee shareholder, and an employee or employees). Such a company will have a single residual claimant and no conflicts between owners. They also assume that because costly mechanisms for separating management and control decisions are not needed, they will not exist. In practice, if the corporate form is used in such a situation, many of these mechanisms will still be present since the legal corporate form requires them regardless of whether such separation actually exists. The costly mechanism is not avoided despite being completely unnecessary. Fama and Jensen assume that their logic is followed both by the workings of the legal system and by those choosing the legal form of their business. This is not necessarily the case.

McNulty (1984) has written that one limitation of the literature on the nature and role of the firm is its failure to recognize the diversity and heterogeneity of this concept and in particular the existence of the one-person firm (which he calls the unitary firm). He argues that the unitary firm

> cannot properly be viewed as a 'system of relationships', as Coase conceptualizes the firm; it is not formed in order to monitor 'team' production, the reason given by Alchian & Desetz for the existence of firms; and its internal operation does not involve 'collective' action, as in Arrow's schema. Yet to the extent to which it exists, it plays a fundamental economic role no less than its larger and more diversified counterparts

Although McNulty is making a general point about firms without distinguishing different legal forms, this observation applies also to the application of the literature to corporations. The contractual analysis of the company falls down in relation to the single-person firm (Paillusseau 1991).[3] As is recognized by the literature, other explanations must be sought for the choice of the corporate form for the very smallest firms than the reduction of agency costs. The agency literature does not support the view that the corporation is necessarily the most efficient legal form for all types of firm or that use of the form *per se* will encourage growth. The assumption that mechanisms existing for one situation will be avoided when the justification for them disappears is not borne out in practice by the operation of the legal system and its use by business owners.

The legal perspective

There is much praise of the limited liability company in the legal literature also and here too there is often emphasis on the larger company which can lead to an equation of limited liability companies with entrepreneurship. Gower, for example, describes the legislative creation of limited liability as a major economic breakthrough; the 'intervention which finally established companies as the major instrument in economic development' (1992: 47). However, there is also strong recognition that there are distinct types of company; that some companies are not formed with a view to raising capital but are simply a 'device for personifying the business' (Gower 1992: 10).

Whilst the large, capital-raising company is seen to be 'economically . . . by far the most important' (Gower 1992) the law must also concern itself with the companies formed by the sole trader or small partnership, since they exist in great numbers. One role of the law will be to provide suitable rules to facilitate the operation of such companies in as efficient manner as possible. However, law is not only concerned with economic efficiency (Farrar *et al.* 1991: 37, Teubner forthcoming). Legal form will not be judged by lawyers only in terms of profitability, survival, growth and employment. They will also be concerned about the effectiveness of a given legal form to regulate relationships between parties. Company law seeks a balance between two roles: the facilitation of business organization and regulation of the relationships of all business owners with co-owners, creditors and the public. It could be argued that these two roles are entirely compatible, since efficient functioning of markets and organizations will require the confidence generated by effective regulation and prevention of uncompensated transfers of risk (by the abuse of corporate form) which may result in costly processes to prevent such transfers. Nevertheless, it is possible that there will be a point at which the pure objective of efficiency will be considered inconsistent with 'legal policy'.[4] The lawyer's central concern with protection of various parties, as an objective separate from the efficient use of resources and wealth maximization, may result in the pre-occupations of lawyers and economists differing in some respects. They may reach similar conclusions from these different perspectives, but they will be interested in different issues along the way.

Thus there are cases when, despite praise of the limited liability company as a stimulus to entrepreneurship and valuable risk taking, the law does not choose to protect small company owners but favours the third parties concerned. So, for example, the principle of separate legal personality may be breached by lifting the 'corporate veil' with the result that some measure of liability may be imposed on the company owner in a situation where he sought to hide behind the company.[5] It has been cogently argued that certain cases of lifting the veil can be justified on economic grounds as a mode of reducing the social cost of limited liability where it gives an incentive to engage in excessively risky activities (Easterbrook and Fischel 1985). Under-capitalized small companies are thus, justifiably, more likely to be subjected to this process than public companies. But this cannot explain why the veil is lifted as against small undercapitalized companies in some cases and yet explicitly allowed to remain in place as a protection in others. Whilst there might be some economic justification for the end result, the dividing line between those considered to be abusing corporate form and those considered to be using it to obtain limited liability or separate legal personality in the way intended is a line which is drawn by the courts based on some notion of legal policy, albeit not always very well articulated.

There is, then, a tension between the enabling and the regulatory functions of company law. The availability of cheap and easy incorporation with limited liability, apparently desirable for the stimulation of the economy, can lead to abuse. In 1962, evidence given by the Board of Trade to the Jenkins Committee referred to 'the irresponsible multiplication of companies', to 'the dangers of abuse through the incorporation with limited liability of very small under-capitalised businesses' as well as to the administrative problems to which they could give rise (DTI 1981: para 1.12). Free access to limited liability is not without its drawbacks to society and to the businesses concerned and some restrictions are necessary.

In the UK, the balance between facilitation of business through easy access to the corporate form and regulation for the protection of third parties has been sought in part through a policy of disclosure. Disclosure is frequently said to be 'the price to be paid for limited liability'. There is a disclosure cost attached to selecting to operate through a limited liability company but, confusingly, it is not charged up front; it may become apparent only after a period in operation after which it begins to be necessary to file returns and produce audited accounts. Such costs may be taken on lightly and without full understanding, later leading to discontent when their true extent is discovered. Other costs of incorporation, relating to the statutory and fiduciary duties of directors, for example, similarly may be under-stood fully only after incorporation or even on breakdown or insolvency of the business. This can mislead the business owner into taking on such requirements without proper consideration.

In recent years, disclosure has been found inadequate to protect the public from those prepared to abuse the privileges of the corporate form and stronger measures to increase the protection of creditors have been introduced. Although no qualifications are required to set up a company, s.214 of the Insolvency Act 1986 increases the likelihood that a company director will become person-ally liable for the debts of the company where he ought to have concluded that there was no reasonable prospect of the company avoiding insolvent liquidation. Provisions for disqualification of directors abusing the corporate form have also been strengthened, in the Company Directors Disqualification Act 1986.

These new provisions are still undergoing judicial development and have been criticized for being poorly drafted and under used (Finch 1990; Cork Gully 1993). In some cases the courts have continued to place great emphasis on the importance of supporting the concept of limited liability. So, for example, in *Re Douglas Construction Services Ltd* v. *Anor* 1988) 4 BCC 553 at p. 557, Harman J. refused to disqualify a director who had mismanaged two companies, stating

> It is of vital importance that the court, in operating this very important jurisdiction created by Parliament for the protection of the public . . . should be careful that it does not so act as to stultify all enterprise. The purpose and the great value of the invention in 1862 of the limited liability company was to enable entrepreneurs to take risks without bankrupting themselves.

On the other hand where two directors (husband and wife) did not act dishonestly, John Weeks QC in *Re DKG Contractors Ltd* ([1990] BCC 903) held them personally liable for the company's debts under s.214 even though

> Neither of them had any knowledge of company law or of the concept of limited liability. Mrs Gibbons did not know to what extent she might be liable for the company's debts. I do not think that they deliberately traded in the manner in which they did in order to avoid personal liability. However, I do not think that they acted reasonably. Before trading in the manner in which they did, they ought to have sought some advice at least.

If this latter approach is developed by the courts, as many lawyers would argue is necessary for the protection of the public, then easy access to limited liability could become a severe trap for the unwary. Whilst it may well be more reasonable to impose such responsibility on unwise business owners than for innocent third parties to suffer, there is nevertheless an element of locking the stable door after the horse has bolted here. Arguably a more honest and valuable policy would be to make incorporation more dependent upon fitness to run a limited liability company in the first place (Hudson 1989). Instead great importance is attached to ease of incorporation, but then much is made of the 'price to be paid' for this 'privilege'. Severe criticism is reserved for those who 'abuse' the privilege, even if they do not understand the nature of the arrangements they have entered into, and gain little from them personally, whilst those who set out to abuse the corporate form may often contrive ways to do so, with the legal penalties available being difficult to enforce.

It is a function of law to regulate forms of business which, whilst not necessarily 'efficient' or of great importance to the wealth creation at a macro level, exist in significant numbers and fulfil a role for their owners. As shown below, there are many one-person and other very small businesses which incorporate for a variety of reasons. In addition, the numerical majority of firms in the UK are unincorporated sole traders or partnerships. VAT figures for 1990 show 34 per cent of VAT units are companies, 26 per cent are partnerships and 38 per cent are sole traders. This almost certainly underestimates the percentage of unincorporated firms in the population as a whole since those below the threshold for compulsory VAT registration will only be included if they have registered voluntarily, and the smaller the firm, the more likely it is to be unincorporated. Hence firms which may be of little interest to those primarily concerned with growth, employment, profitability and survival must be one focus of attention for lawyers. Sensible regulation must be provided for them at a level appropriate to the nature of the organization. It should facilitate growth if that is an objective of the owners, but it should also facilitate the smooth operation of the non-growth firm which is providing a function important to its owners and needs a framework which is 'efficient' within their terms. Unless there is a clear policy reason not to provide for such businesses, it is the task of the law to provide an enabling framework. If the legal form provided does not prevent abuse or is unduly burdensome, the law is not performing its role. Subject to this, however, the law should attempt not to distort commercial behaviour in such a way as to reduce efficiency. At this point the interests of economists and lawyers will converge.

The legal texts investigate thoroughly the single-person company. In his leading text, Gower 1992: ch 5) commences his discussion of the advantages and disadvantages of incorporation with a passage on *Salomon* v. *Salomon & Co* [1897] AC 22, HL, the case which finally established the legality of the 'one-man' company and showed that incorporation was as readily available to the small private partnership and sole trader as to the large public company, but it also revealed that it was possible for a trader not merely to limit his liability to the money which he put into the enterprise, but even to avoid any serious risk to the major part of that by subscribing for debentures rather than shares.

Gower then sets out the traditional legal reasons for incorporation: limited liability; corporate personality which enables the company to hold property in its own name, to sue and be sued and gives perpetual succession; transferability of shares and the availability of the floating charge (a flexible form of finance, not available to unincorporated firms for technical reasons allowing stock-in trade and book debts to be used as security without restricting use of that stock). Taxation is referred to: under the current regime, tax and national insurance considerations may militate against incorporation, but there are also circumstances in which incorporation brings tax advantages. Disadvantages of incorporation for the small business are generally seen to be the level of formality and publicity (for example, the need to maintain and file audited accounts and complexities if the business is to cease to exist) and rules restricting free withdrawal of capital from limited companies.

In theory, a number of the advantages listed by Gower are of value to the very smallest firms. Unrestricted transferability of shares is, however, not a feature of private companies: in an owner-managed firm, the identity of shareholders will be too important to the original shareholders to permit their co-owners such freedom. Limited liability is also often said to be largely illusory in the case of small private companies because of the operation of the market which is likely to result in lenders insisting on personal guarantees from owner-managers. However, limited liability will protect small company owners from liability to general trade creditors who will not normally secure personal guarantees and sometimes in respect of liability for negligence and other torts.

Recent legal literature, especially that developed in the USA, has been heavily influenced by the economists' model of the company as a device for the reduction of monitoring costs (Easterbrook and Fischel 1986; Posner 1986, and the literature cited there). The legal literature produced by this school distinguishes carefully between public and closely held corporations and recognizes that the factors leading to incorporation by such firms are not identical and that the regimes applicable to them cannot sensibly be identical (Manne

1967).[6] It has been convincingly argued by Halpern, Trebilcock and
Turnbull (1980) that an unlimited liability regime would be the most
efficient for small, tightly held companies for whom a limited liability
regime creates a moral hazard and transfers uncompensated business
risks to creditors, thus inducing costly attempts by creditors to reduce
these risks.[7] This argument ties together regulatory and efficiency
concerns and legal and economic perspectives.

Sometimes, the operation of the market does achieve something
approaching an unlimited liability regime for *incorporated firms*, in
circumstances where limited liability creates an unacceptable risk for
creditors (at least this is achieved by those third parties in a strong
bargaining position). In such cases, the law does not seek to uphold
the principle of limited liability in the face of market pressure. Thus,
the banks are not restrained by law from taking personal guarantees
from company directors. Landlords and major trade creditors also
can and do require personal guarantees. The argument that the
limited liability company is not always an appropriate legal form is
therefore supported empirically as far as relationships with third
parties are concerned. In the project described below, the authors
set out to gather further empirical evidence of the suitability of the
corporate form for very small businesses.

EMPIRICAL EVIDENCE

The economic and legal literature discussed above puts forward
factors which will influence business owners to choose the corporate
form. It suggests that the corporate form will not prove worthwhile
for very small firms which will not be able to obtain the benefits of
limited liability in practice and which have no large demands for
unrestricted risk sharing and specialized decision skills and no large
demands for wealth from residual claimants. In essence the issue is
one of setting the costs of formalities, complexities in the applicable
rules, restrictions and publicity against the benefits which are obtained
on incorporation. It is this cost/benefit balance which the authors set
out to investigate empirically.

To explore the theme that the costs of incorporation may sometimes exceed the benefits we use the reasons given by small business
owners for choosing to incorporate or to be unincorporated and
further analysis of the associated data collected in the course of the
same project to investigate the nature of the limited liability company
as it exists within the UK small business community. It will be seen
that the reasons given for incorporation by the survey respondents

are not always those which have been hypothesized in economic and legal theory. This may mean that analyses based on the supposedly 'rational' reasons for incorporation and the consequent conceptions about the relative populations of the incorporated and unincorporated business communities proceed on a mistaken basis.

The survey

The data described below form part of the responses to a mail questionnaire survey of sole traders, partnerships and owners of small limited companies supplemented by telephone and face to face interviews with selected respondents and auditors. Information was obtained from 429 firms in four geographical areas during 1990 and 1991. The overall objective of the project was to assess the level of satisfaction amongst the business community with the legal forms available to small business owners and to investigate the effects of differing tax regimes applying to these different forms. Respondents were questioned about the general characteristics of the business, reasons for choice of legal form, disadvantages of that form, sources of finance, and related topics. The project is further described in the Methodological Appendix (see p. 275). For a fuller examination of a wider range of the data collected see Freedman and Godwin (1992).

The figures given here are based on the responses to postal questionnaires from 125 small limited companies, except where otherwise specified. Some results are compared with findings from parallel questionnaires completed by 146 unincorporated firms. Percentages given below are percentages of those eligible to reply, including 'don't knows' and non-responses, unless specified otherwise.

Sample selection method

The unincorporated firms surveyed were selected at random from the Yellow Pages telephone directory, it being clear that the vast majority of unincorporated firms are very small (see Figure 7.1) and there being no comprehensive register of unincorporated businesses. The companies were selected at random on a post code basis from the Companies Register, using a commercial intermediary (Jordans) to assist with selection. Although the use of the Companies Register reduced the response rate, since it is not entirely up to date and contains many companies which are not trading, it had the advantage of allowing us to eliminate companies known to have a turnover of over £1 million. Turnover is not known for companies filing modified

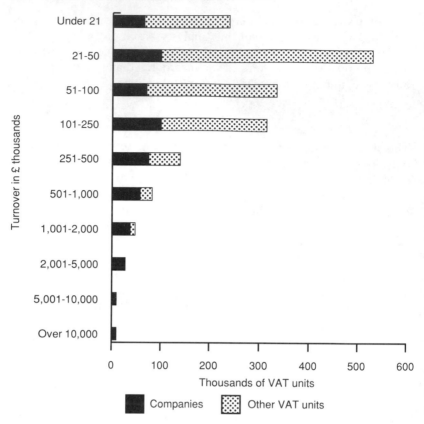

Figure 7.1   Companies and unincorported firms registered for VAT, 1990

(now known as 'abbreviated') accounts – an option for small companies within the Companies Act definition – or with no filed accounts, such as new companies. These were included in the sample although it was possible their turnover exceeded our limit. Use of the Companies Register rather than a telephone directory also ensured that our sample included the very smallest firms and others not serving the public but nevertheless trading and using the corporate form. These proved to be of great interest, vindicating our sample selection method.

Size of respondent companies

Ninety-three (79 per cent) of our respondent companies had a turnover of £1 million or below (after elimination of those not replying to this question). Twenty per cent had no employees (other than the director/owners) and 54 per cent had five or fewer employees (the median number was four). Eighty-four per cent of our respondent companies had only shareholders who were also directors and 68.4 per cent had only two ordinary shareholders. Ninety-six point five per cent had five or fewer ordinary shareholders. Many were husband and wife companies; it is probable that in many cases a spouse was brought in as a second shareholder to meet the two shareholder requirement which was not removed from UK company law until 1992.[8]

These were precisely the companies we were targeting; most were not obviously using the corporation as a capital raising device. We can formulate a working definition for our purposes of a micro company as a company with a turnover of £1 million or less,[9] five or fewer ordinary shareholders and twenty or fewer employees. This would cover 72 per cent of those respondent companies for which we have such information.[10] It is amongst these companies that we are likely to find use of the corporate form where theory would not predict that this would be useful. Many of these entities will not wish to, and will not, grow (Gray 1992). However, this group will also include entities which are formed with an intention of growth and which will ultimately wish to use the corporation as a capital raising device for which limited liability will be essential. Clearly it is difficult to distinguish these types of micro company, using objective measures, at start up. However, such a distinction would be very useful, not only in the design of appropriate legal structures and advice on which available package to choose, but also for lenders and others dealing with the enterprise.

Reasons for incorporation

Respondents were given a list of thirteen possible advantages of incorporation, and were asked to indicate how many were important to them; additional reasons could also be written in. As expected, limited liability was considered important by a majority of our incorporated respondents, although not universally so (see Figure 7.2). Those referring to limited liability as important broke down as follows: limited liability to suppliers important (46.4 per cent), limited liability to banks and financiers (45.6 per cent), limited liability to customers (33.6 per cent), limited liability to other shareholders (16.8 per cent). As these are overlapping groups, the overall proportion stating that limited liability of some kind was

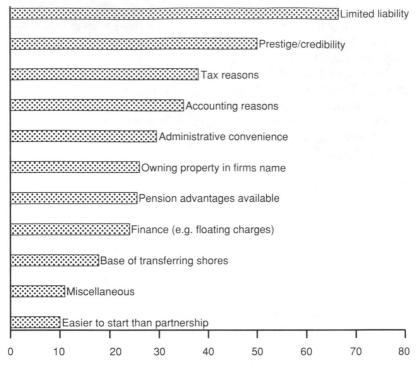

Figure 7.2   Reasons for incorporating: % of all respondent companies

important was two-thirds (66.4 per cent). One in seven (14.4 per cent) definitely stated that limited liability was not important and the remainder (19.2 per cent) did not make any response relating to limited liability.

Although limited liability to banks and financiers was widely mentioned as important it is often argued, as discussed above, that such limited liability is illusory because these creditors so often insist on personal guarantees. Of our sample, 53.6 per cent (67 respondents) stated that directors and/or their spouses currently provide personal guarantees to banks or other creditors. Where further information was supplied, it seemed that these guarantees were almost always given to banks.[11] Of the sixty-seven companies where such personal guarantees were given, almost half (thirty-three) stated that limited liability to banks and financiers was an important reason for incorporation. For these company owners, the limited liability obtained by incorporation is, at best, incomplete (see Binks 1991).[12]

Other than limited liability, the reason most often mentioned as important was prestige and credibility (50.4 per cent). This was well ahead of tax reasons (38.4 per cent), accounting reasons (35.2 per cent), administrative convenience (29.6 per cent), ownership of property in firm's name (26.4 per cent) and availability of pension advantages (25.6 per cent). Ranking after these factors were raising finance (e.g. floating charges, seeking outside equity finance) – (24 per cent) and ease of transferring shares (17.6 per cent). Thirteen respondents thought that starting a company was easier than starting a partnership; although this is not a large response (just over 10 per cent of respondent companies), it is of interest in view of the conventional wisdom that starting an unincorporated firm is simpler and cheaper than incorporation.

It is widely recognized that incorporation is seen as bestowing prestige and credibility, although when limited liability companies first appeared it was thought that this very limitation on their liability would result in mistrust.[13] It is ironic, then, that a notion seems to have arisen in the commercial world that incorporated firms are somehow more serious and trustworthy than unincorporated ones. This was borne out by the responses to a question on attitudes to companies: around half believed that banks (52 per cent), suppliers (51.2 per cent) and customers (48 per cent) preferred to do business with limited companies. The obligation to disclose information on the Companies Register may account for some of the confidence shown in incorporated firms – although only 4.8 per cent of respondent companies often made use of the accounts of other companies filed at Companies House. The simple fact of registration and consequent increased visibility may also help. However, the real explanation may be linked with the fact that the bigger the firm, the more likely it is to be incorporated – see Figure 7.1.[14] Overall, this may lead to the impression that a company is more reliable than an unincorporated firm. Given the large number of very small companies with little or no capital, as discussed below, this is clearly not a valid universal assumption. Incorporation is not an indicator of size or standing.

Taxation

The self-employed pay tax under Schedule D of the Income and Corporation Taxes Act 1988 currently on a preceding year basis. Tax is payable twice yearly at the prevailing rate of income tax for individuals, and is based on the profits of an earlier period. On incorporation, even within a one-person business, the directors' and employees' remuneration will be charged to tax under Schedule E and tax will be collected from the company monthly or quarterly under the PAYE system on a current year basis. The company will be subject to corporation tax on its profits net of any such remuneration, but for companies with profits of £250,000 and below this is set at a special small companies rate of 25 per cent.[15] Dividends are taxable in the hands of the recipient and not deductible from profits but the shareholder recipient receives a tax credit for tax paid by the company. This *imputation* system prevents double taxation of dividends to a considerable extent.[16] Profits retained in a company paying the small companies' rate will be sheltered from the higher rates of income tax.

Forty-eight companies (38.4 per cent of respondents) stated that tax was an important reason for incorporation. This sub-group was asked some detailed questions on tax reasons. The reason mentioned most often was income tax/corporation tax rates (thirty-five firms): the small companies' corporation tax rate of 25 per cent may become attractive once the individual taxpayer's income moves into the higher rate band, although this has to be weighed against disadvantages such as higher National Insurance payments on directors' remuneration than on that of a self-employed trader. In some cases, such external evidence as we had tended to suggest that incorporation might not be the best option for the firm concerned in current conditions,[17] but the reason may nevertheless have been based on good advice at a time when income tax rates were higher and the tax factors were differently balanced. The tax cost and company law complexities of disincorporation can result in some companies becoming trapped in incorporated form even if they become aware that this is no longer beneficial, whilst others simply may not be aware of the shift in advantage.

Twenty-two firms referred to timing of payments. Although it is usually thought that there is a cash flow advantage in paying on a preceding year basis this can have disadvantages. It requires the retention of reserves and can cause problems, especially if profits are in decline and tax has to be paid on the profits of a good year in a bad one. Some business owners apparently prefer to pay as they earn. It was announced in the March 1993 Budget speech that the self-employed are to be taxed on a current year basis as from 1996–7 (IR 1991; IR 1992). This is welcome as it will remove one complexity and a distorting factor as regards choice of legal form.

Twenty-two firms mentioned capital gains tax planning and sixteen referred to inheritance tax. Eight companies referred to the business expansion scheme (which terminated at the end of 1993) as an important reason for incorporating.

Significance of perceived advantages

Analysis of these responses gives rise to doubts about the adequacy of the theoretical accounts of the reasoning underlying choice of legal form. Our survey evidence suggests that although the predicted economic or legal reasons for using the limited liability company are important, some, such as raising finance and transferring shares, are perhaps less significant than had been anticipated. In addition, other factors – which we might call 'cultural reasons' – must also be taken into account. These can be summed up in terms of the prestige which companies are perceived to have. Such reasons seem to have been encouraged by policy makers in promoting the 'enterprise culture'. As the idea that companies have greater prestige than unincorporated firms appears to be pervasive within the business community, a choice based on this factor is not 'irrational' or unreal but we would expect that ultimately, as the costs of incorporation became apparent, those who incorporated mainly for such cultural reasons would be less satisfied with the corporate form than those incorporating primarily for what we have called here 'economic or legal reasons' for whom the costs would be more likely to be acceptable as conferring some benefit. Misconceptions about legal form may result in some small and simple firms taking on responsibilities, duties and procedures more appropriate to larger and more complex firms; and thus they may find themselves operating less efficiently and economically than they might otherwise have done. Furthermore, firms which realize that incorporating was a mistake may find that they have become trapped in company form due to the difficulty and expense of disincorporation. These hypotheses are explored further in the next section.

Taxation as a reason for incorporating falls into a category of its own. It is predictable that tax differences in the treatment of different legal forms will result in choice of legal form for tax reasons. It is therefore not surprising to find empirically that this is so in nearly 40 per cent of cases. Rational though this may be as a choice on the part of incorporators, however, it suggests an irrationality in the system, in the sense that it may distort choices which could otherwise be based on factors linked to commercial suitability of a structure as an efficient method of organizing resources. Unless there are sound policy reasons for wishing to use the tax system as an incentive to incorporate or not to do so, any tax differences between the two structures may introduce an inefficiency. This will be worsened by changes in the tax system where it is difficult to change form to meet altered conditions. Therefore, we would also expect certain groups of company owners who incorporated for tax reasons to be dissatisfied with incorporation as a method of organizing their business. Again, this is explored in the next section.

Disadvantages of incorporation

The possibility that for some companies there are disadvantages which outweigh the advantages of incorporation was explored in the questionnaire: firms were asked whether the advantages of incorporation outweighed the disadvantages, and were invited to indicate whether certain often-cited disadvantages were important to them.

Respondents were given a list of seven possible disadvantages of incorporation, and were asked to indicate how many were important to them (see Figure 7.3); additional reasons could also be written in. Seventy-two per cent of respondent companies agreed or agreed strongly that the cost of statutory audit was too high. This was far higher than the number agreeing that administrative burdens were a problem of incorporation (52 per cent), lack of confidentiality (50.4 per cent), expense and complexity on setting up (33.6 per cent) and the need for two shareholders (32 per cent); loss of personal control and loss of control over capital were thought to be disadvantages by only fourteen firms (11.2 per cent) in each case. It was also higher than any of the disadvantages reported by unincorporated firms. The resentment of the statutory audit was clearly linked to profit levels. Of those providing the relevant information, over 80 per cent of those with profits below £100,000 agreed that these costs were a disadvantage of being incorporated, but this dropped below 50 per cent in relation to companies with profits between £100,000 and £250,000.

This strong response on the statutory audit has since been confirmed in interviews with companies and auditors and has been discussed by the authors elsewhere (Freedman and Godwin 1993). These findings coincided with various pressures on the Government to review its policy of mandatory audit for small firms, not least from the Institute of Chartered Accountants (ICAEW 1992). The Chancellor's March 1993 Budget speech responded to this pressure by announcing the forthcoming publication of a consultative document 'setting out options for reducing this burden' and accepting that the 'current statutory audit requirement imposes a disproportionate

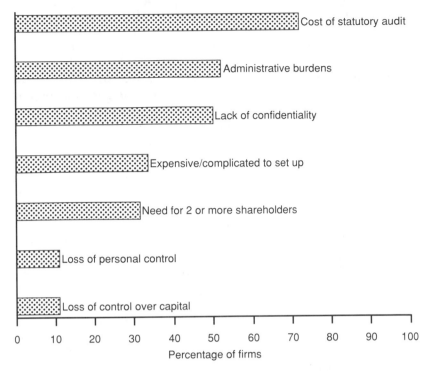

Figure 7.3   Disadvantages of incorporating: % of all respondent companies

cost on smaller businesses'. The audit requirement is an example of a costly mechanism intended primarily for the situation where management and ownership are separated. When imposed where this is not so, it becomes an unwarranted cost of the incorporation mechanism with no benefits apparent to the business owners. However, it cannot be avoided or ignored as it is a statutory requirement and so is a genuine burden on businesses. The debate on the abolition of the mandatory audit requirement has also illustrated clearly the difficulties of defining the companies to whom reforms are to apply. Much opposition to change undoubtedly results from the very different impressions of the type of company which would be affected. If all companies within the Companies Act small company definition were to be excluded from the requirement, this would not only cover the vast majority of companies in the UK but would also relieve from the audit some companies with very substantial businesses, minority interests and a number of employees (see Statistical Appendix, p. 270). Lack of statistical information about very small companies makes it difficult to reach agreement on a suitable threshold (ICAEW 1993).

By contrast, in practice, other, more internal, requirements, such as that to have annual meetings, can simply be ignored by owner-managed companies as performance can be perfunctory or fabricated and cannot be monitored effectively: about 30 per cent of our respondents seldom or never held shareholders' meetings for example. Therefore complex or unsuitable rules do not appear always to impose compliance costs. It follows that provisions designed to give relief from such supposedly burdensome rules will not necessarily be particularly helpful. Interestingly, when asked whether various deregulation measures in the 1989 Companies Act would help them, respondents were lukewarm in their welcome of provisions enabling them to dispense with the annual general meeting (38 per cent thinking it would make no difference and 46 per cent thinking it would be helpful) whereas 74 per cent thought that an administrative change simplifying the annual return (a statutory form which must be filed at the Companies Registry) would be helpful.

Over one-fifth of companies responding to the questionnaire (27 companies – 21.6 per cent) considered that the disadvantages of their chosen legal form outweighed the advantages, compared with only 4.8 per cent of unincorporated firms.[18] This does not necessarily indicate that the incorporated form is less likely to be satisfactory than operating as a sole trader or partnership: the explanation may be that it is simpler for an unincorporated firm to change form if it wishes to, so fewer dissatisfied unincorporated firms are likely to remain in that form. This explanation is borne out by the fact that 40 per cent of the company respondents had traded previously in unincorporated form, whereas only two of the unincorporated firms had previously been incorporated. It does however illustrate the danger of becoming trapped in corporate form – problems associated with ill-advised incorporation are likely to persist, whereas a firm which starts up inappropriately in unincorporated form can take steps to remedy the situation.

Of these dissatisfied companies, only five had any plans to change legal form. The reasons given for not changing were tax obstacles (seven companies), that the alternatives are just as bad (five firms), that the respondent did not know it was possible (four firms), that disincorporation is too complicated (five firms) or too expensive (five firms), and difficulties with creditors (two firms). In addition, one of the five planning to change legal form was in fact ceasing trading completely and he did not know it was possible to disincorporate.

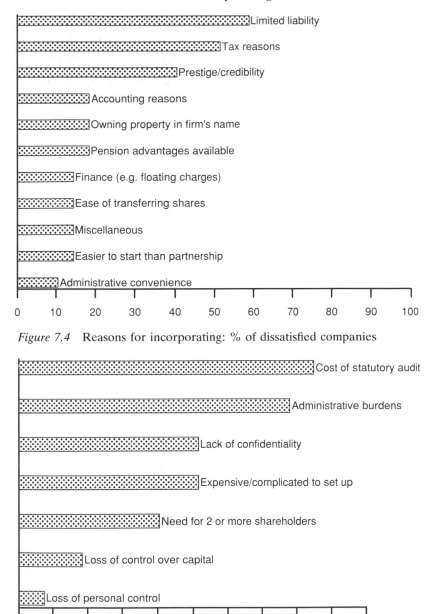

Figure 7.4   Reasons for incorporating: % of dissatisfied companies

Figure 7.5   Disadvantages of incorporating: % of dissatisfied companies

Two companies which were in the process of disincorporating also commented on the tax obstacles to this process, as did one respondent who did not answer the question on plans to change. Thus as many as ten of the twenty-seven dissatisfied companies were concerned about the tax obstacles to disincorporation. This suggests that, despite difficulties encountered with earlier attempts at facilitating disincorporation, Government should persevere with reform and that the key will lie in finding a sufficiently simple form of tax relief for these very small firms. The DTI is to examine this issue as part of its forthcoming review of company law (DTI 1992a). If relief on disincorporation could be targeted at micro companies it is possible that some of the complexities which have arisen previously could be avoided.

The reasons for incorporation of the twenty-seven dissatisfied companies were compared with the reasons of the companies overall; results are shown in Figure 7.4. Differences between the dissatisfied companies and others were marked. *Limited liability* was only given as a reason for incorporating by 59.3 per cent of the dissatisfied companies (against 66.4 per cent overall) *accounting reasons* by 18.5 per cent (35.2 per cent overall), *obtaining finance* by 14.8 per cent (24.0 per cent overall) and *administrative convenience* by 11.1 per cent (29.6 per cent overall). Thus their owners do not, for whatever reason, consider the theoretical attractions of incorporation as important as do other company owners. Dissatisfied companies are more likely to have incorporated for *tax reasons*: 51.9 per cent of dissatisfied companies mentioned tax, as against 38.4 per cent overall. They are also more likely to perceive disadvantages (Figure 7.5). Differences between the dissatisfied companies and others answering the questions on disadvantages showed that the former resented the audit more (85.1 per cent compared with 72 per cent overall), complained more about administrative burdens (77.7 per cent against 52 per cent overall), saw incorporation as complex 51.8 per cent against 33.6 per cent overall), more often found the need for two shareholders burdensome (40.7 per cent against 32 per cent overall) and experienced a loss of control over capital (18.5 per cent against 11.2 per cent overall). The smallest firms were the most likely to be dissatisfied with the corporate form; 36.8 per cent of those with a turnover below £100,000 but only 10.1 per cent of those with a higher turnover fell within this category.

This is consistent with our hypothesis: where choice of form is not based primarily on what we have called above 'legal and economic' considerations, inefficiencies will arise. The data suggest that people who are incorporating mainly for prestige, tax and other non-commercial reasons are more likely to be dissatisfied with the corporate form than others. They may not gain the advantages they had hoped for and, even if they do obtain some advantage, the audit cost, and other administrative burdens such as the annual return, may exceed the benefit for them. However, requirements which are internal and cannot be externally monitored may not impose costs in that they may be ignored.

Employed or self-employed? On the frontier of the small business sector

During the analysis of dissatisfaction, a small but distinct group of workers was identified. Six respondent companies, all providing business services (four computer consultants, a surveyor and a draughtsman) gave tax as their main or only reason for incorporation. Each obtained work from only one or a few companies, or by offering services through an agency. Four of these six felt that the disadvantages of incorporation outweighed the advantages, and the other two reported some drawbacks to incorporation. All six complained of the cost of the statutory audit and only one said that the information prepared for filing was useful (and then that only 'some of the information is useful'). Checks on the Companies Register showed that all these companies had two directors (husband and wife in each case) who owned all the shares and there were no other employees.

Prima facie the corporate form does not seem appropriate to this situation. In many respects these workers are akin to employees. They provide no equipment, employ no staff and must do the work for which they are engaged themselves. On the other hand, they are not entitled to holiday or sick pay or other employee benefits. Since they have some of the characteristics of employees and some of the self-employed, we describe them as 'hybrid workers'. As a result of this hybrid nature, one-person service businesses operating in unincorporated form run the risk that the Inland Revenue will argue that they are employees, subject to deduction of tax at source under PAYE, rather than self-employed, taxable under Schedule D. Payments to self-employed workers (other than sub-contactors in the construction industry and those working through agencies) may be made gross, which may give a cash flow benefit, and there is greater scope for deduction of expenses from such fees than from employees' remuneration. Possibly even more important, National Insurance payments for the self employed are substantially lower than for employees. (There are also differences in benefit entitlement but these are not commensurate.) The law on the distinction is dependent on decided cases and to a considerable extent a question of fact.[19] This makes it difficult for an individual to contest an Inland Revenue classification, although not impossible. The Inland Revenue has recently lost one such case in which it was decided that it is possible for a person to be self employed even if all he does is supply his own services and he has no risk of loss (*Hall* v. *Lorimer* 1992 STC 599, confirmed by the Court of Appeal 1994 STC 23). This is just one aspect of the difficulty produced by the legal employee/self-employed distinction (see Hakim 1988 on the statistical problems); a classification which is inadequate to deal with the many different labour relationships arising in current economic conditions.

Those working through agencies are treated as employees by statute unless certain exceptions apply (s.134 Income and Corporation Taxes Act 1988). This legislation gives implicit recognition to the fact that the PAYE system is used as an anti-fraud device. However, it has the presumably unforseen consequence of pushing people into a totally inappropriate legal form. If they had not formed companies, our 'hybrid' respondents would have been treated as employees for tax purposes either under the case law or under the agency rules, in which case each company using their services would have to take on the cost and administrative burden of paying tax at source under the PAYE system. Our survey revealed that agencies often insist that their clients incorporate to avoid this administrative burden falling on the agency. If the business is incorporated, the agency rules do not apply and it is much more difficult for the Inland Revenue to argue that there is not a separate business. So far, attacks by the Inland Revenue on such arrangements as artificial have not succeeded (Gable 1989; *Inland Revenue Tax Bulletin*, February 1992). Therefore, on incorporation, the company receives payments gross so that the person to whom services are supplied is not involved in operating PAYE. Although tax must be collected under PAYE on any remuneration paid to directors by the one-person company, not all the company's income needs to be paid out in this way; some may be retained and sheltered from higher rates of tax or it may be paid out by way of dividend with a consequent saving on National Insurance. Expenses may be deductible by the company which an employee would not be permitted to deduct, although in some cases these will then be assessed as taxable benefits in kind on the owner/director so that little will be achieved.

Sometimes this type of arrangement is the choice of the worker, who may achieve a tax advantage, as was highlighted in relation to

John Birt's work at the BBC (Boggan, The *Independent on Sunday*, 7 March 1993). As some news stories suggested in Mr Birt's case, the total saving may not be very significant and part of the reason for this will be the costs of running the company, especially the statutory audit. This will not be a matter of great concern to those designing the legal system since such users of the corporate form are generally well advised and aware of the costs and benefits. They have a real choice as to whether to use this form and will be aware from the outset that this is not the situation for which this legal form was actually intended. However, our respondent hybrid workers did not seem to be in this category. It was not clear that they were engaged in tax planning, at least not on their own behalf. Most appeared to have no real choice. They all agreed that their customers preferred them to be, or even insisted on their being, incorporated. In fact they felt that they had been forced by the labour market to incorporate: they could not obtain suitable work in any other way.

A study of these respondents' replies, as well as telephone and anecdotal evidence, written comment[20] and advertisements in journals placed by company formation agencies suggest that self-employed persons are being encouraged to incorporate by the large companies which use their services.[21] These large companies then achieve savings, not only as described above in the form of reduced administration in not having to operate PAYE but also, perhaps, by escaping employment legislation and passing on labour costs to the workforce in other ways (Collins 1990; Rainbird 1991). Leaving aside the merits or otherwise of this behaviour, it would seem from the literature that there are also less contentious commercial reasons for the use of a flexible workforce in this way (Wood and Smith 1989; McGregor and Sproull 1992). Provision of specialist skills and the matching of manning levels to peak demand are far the most common reasons given by establishments for use of self-employed workers. This was borne out by our interviews. For example, one hybrid worker told us that the petro-chemical industry could not survive without 'floaters' because certain skills were required only at certain times during the life of an oil rig. For the purposes of this discussion, the point is that the corporate form is chosen in this type of case not by the individual worker but by the industry being served. To some extent tax legislation and tax uncertainties push the large firms in this direction. The individuals for whom the inefficiency of the legal form is most apparent are not those who have chosen it and therefore their resentment of the corporate form is reasonable and predictable. These workers appear to blame tax law rather than company law for their predicament. One commented 'Unfortunately it would be the laws governing Schedule D which need reforming to enable me and 500,000 people like me to operate in a sensible way'.

The answer does no doubt lie in a reform of tax and National Insurance law and, in particular of the archaic Schedular system. However, the examination of this small but significant group is of interest in the context of this chapter since they are identifiably dissatisfied as would be predicted from the fact that their choice of legal form has been distorted by tax considerations. The 'genuine commercial benefits' of incorporation such as limited liability are not of interest to these business owners, so that the burdens fall heavily.

Breaking down the corporate classification

The data analysed so far suggest that very small firms are incorporating for a wide variety of reasons. These we have described as 'legal and economic reasons' on the one hand (those which would have been predicted from the theoretical literature) and 'cultural' and tax reasons on the other. We have put forward a working definition of the micro company based on the characteristics of the bulk of our respondents. This definition is considerably narrower than the small company of the Companies Act or of most economists' definitions. However, it seems that further helpful sub-divisions could be made even within this group. There is a serious lack of hard information about companies at this very small end of the scale, not least because small companies under the Companies Act definition need only file abbreviated accounts which limits the information available as described in the Statistical Appendix (see p. 270). However, a further breakdown of this group would be valuable for the purposes of those formulating policy for and working within the small business sector. In view of the absence of statistical information, further empirical work on a large scale would be needed to explore this in more depth. Here we make a tentative suggestion based on our data that there are objective criteria which could be used to break down this group yet further.

Capital as a distinguishing factor

The difficulty is to distinguish companies formed for 'legal and economic reasons' (economic companies) for whom the benefits of the corporate form are likely to outweigh the burdens, from the companies formed for 'cultural' reasons or distorting tax reasons, who are less likely to find this legal form valuable in the long run. Size alone is not an indicator of this distinction because a company may be very small on start up, but it may intend to grow and actually do so. It might then make sense for it to be incorporated from the outset, although there are good reasons to start as an unincorporated firm and incorporate at a later date. Although we have no data on growth, it is a reasonable hypothesis that the economic companies are more likely to be potential growth companies. The others will often be essentially 'own account workers' or little more and may wish to stay that way. All are subsumed within the one legal category. Since the subjective reasons given for incorporating overlap, with the majority giving limited liability as a reason, it is desirable to seek a more objective method of differentiating these groups.

If a major theoretical reason for incorporating is to raise capital (Posner 1986; King 1977) we might suppose that companies would consider share capital to be an important source of finance. This was not borne out by our survey results which, as already stated, showed that raising finance was not a primary reason for incorporating (Figure 7.2, p. 246). We therefore explored this further.

Asked about sources of finance at foundation (see Figure 7.6), *unincorporated* firms frequently (70.5 per cent) gave capital contributed by the owners as an important source. Bank borrowing was important to 55.5 per cent and trade creditors to 51.4 per cent of unincorporated businesses. Retained profits were also significant (45.9 per cent); 'at foundation' here is presumably being read by respondents as meaning during the early days of the business. This adds up to a picture of owners of unincorporated businesses, the majority of whom are making contributions of capital to their businesses, which, in their eyes at least, are not trivial.[22]

In sharp contrast to the 70.5 per cent unincorporated firms stating that owners' capital was important, only 21.6 per cent of *limited companies* (see Figure 7.7) reported that share capital was an important source of finance on incorporation. Incorporated firms referred to bank borrowing most often (59.2 per cent), followed by trade creditors (50.4 per cent) and retained profits (36.8 per cent). The low response on use of share capital suggests that the formality of contributing share capital and the difficulty of retrieving it discourages contributions, and thus encourages the establishment of undercapitalized small businesses. Only fourteen of the companies commenting on the disadvantages of incorporation thought that loss of control over capital was a disadvantage (see Figure 7.3, p. 251); indeed seventy-seven stated definitely that it was not a disadvantage:

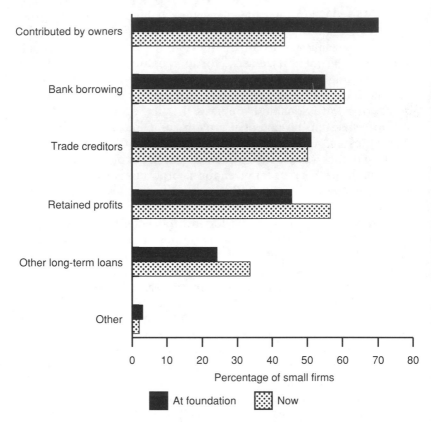

Figure 7.6   Sources of finance for unincorporated firms

but this could have been because company founders had not found it necessary to contribute capital to the company by way of share capital in the first place. Neither did the number of incorporated firms considering share capital to be important increase when asked about their current position; in fact it decreased from twenty-seven to twenty-three firms.

It would not be surprising within a limited liability regime with no minimum capital requirement to find that directors prefer to lend, for example, rather than contribute share capital.[23] However, overall, long-term loans and debentures were not frequently stated to be an important source of finance for respondent companies (19.2 per cent of companies stated that long-term loans were important and 10.4 per cent stated that debentures were important on incorporation). Owners of unincorporated firms cannot, of course, lend money

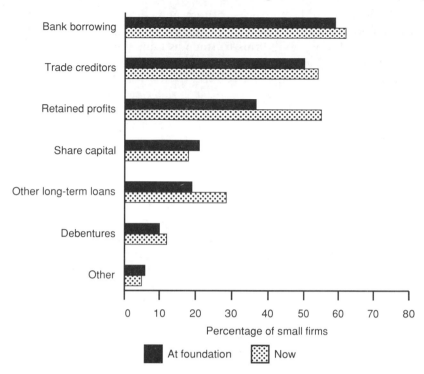

Figure 7.7 Sources of finance for limited companies

to their firms since the firm is not a separate legal person. Of the unincorporated firms, 15.8 per cent stated that long-term loans were important on foundation.

It could be argued that the reason for the greater capital contribution by unincorporated firms is that they cannot raise finance as easily as incorporated firms and so are obliged to contribute more capital themselves. It is often suggested that the inability of an unincorporated firm to give a floating charge might account in part for this difficulty as well as size and lack of sophistication. However, our data show bank borrowing, for example, as important to about the same percentage of firms irrespective of legal form (55.5 per cent of unincorporated firms and 59.2 per cent of incorporated firms considered bank borrowing at the start important). It is nevertheless notable that *more* of the unincorporated firms than the companies stated that bank borrowing was *very* important (41.1 per cent as compared with 36 per cent).[24] It would seem, therefore, that incorporation might actually result in *less* capital being invested in a business than will be the case with an unincorporated business and this supports fears expressed about undercapitalization and the moral hazard of allowing small firms to trade as limited liability companies (Halpern *et al.* 1980; Jenkins 1962).

On the other hand, whatever the reason for the limited capital contributions of company owners, it is not clear that incorporation does in fact protect their personal assets and wealth. As discussed above, more than half of the respondent companies stated that directors and/or their spouses provided personal guarantees to banks or other creditors. To the extent that such guarantees are given, those who borrow in order to start their businesses are putting their capital at risk as much as those who contribute cash. The apparent difference between the financing of incorporated and unincorporated businesses could, therefore, be largely one of perception, but nevertheless of practical importance in terms of attitude to the business. It is possible that at the same time, the system is encouraging some company owners to believe that they are protected when they are not and discouraging the direct investment of capital into the business. This has the further consequence that banks might be protected on corporate failure where they have security, but general creditors will not be, so that this method of allocating risk could have a cost to the business community generally.

Further analysis based on share capital

In order to investigate capitalization of companies in more detail, we analysed the results for incorporated firms by splitting those stating that share capital was an important source of finance at start from those for whom this was not important. For ease of reference, these two groups will be referred to as those with share capital and those without, although the question posed was whether this source was important or not. Thus, some of those described as 'without share capital' may have some such finance, but it will not be significant. This analysis has produced some interesting results, suggesting that the use of share capital may either be a key characteristic, or relate closely to a key characteristic which distinguishes the economic, potentially substantial, entrepreneurial small company from the less growth-oriented micro company (see Figures 7.8 and 7.9). The differences did not appear be explained by sector.[25]

Twenty-seven of the 125 company respondents stated that share capital was important at incorporation. The share capital question in the survey did not differentiate between sources of share capital;

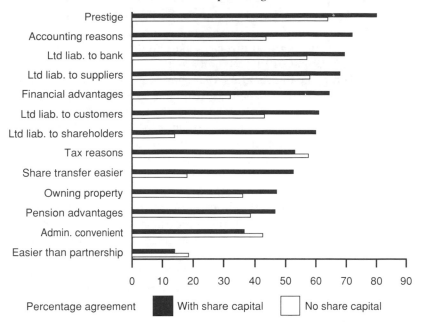

Figure 7.8 Advantages of incorporation, by use of share capital at start

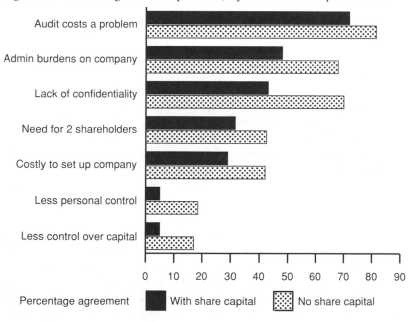

Figure 7.9 Disadvantages of incorporation, by use of share capital at start

however from the answers to other questions it was possible to ascertain that twenty of the company respondents overall (16 per cent) reported having non-director shareholders at the date of the questionnaire. Eight of these twenty were members of the group of twenty-seven companies stating that share capital was important on incorporation; that is 29.6 per cent of this group had non-director shareholders. Even within this group, then, companies were predominantly owned by their managers and the most likely source of share capital at incorporation was from directors.

Respondents reporting *current* high turnover were most likely to have had share capital *at incorporation*: the £¼m–1m band reported the highest percentage with starting share capital (34 per cent), whereas the £10,000–50,000 band reported the lowest (6 per cent). Those considering share capital important at incorporation also tended to have more directors: 3.1 on average (2.7 executive directors + 0.4 non-executives), compared to 2.5 for others (2.0 executive directors + 0.5 non-executives), and more shareholders (3.2 compared with 2.6 for others). In other ways they appeared to have a more substantial corporate structure: for example, 41.2 per cent of those without starting share capital seldom or never held shareholders' meetings, whereas only 12.5 per cent of those with share capital said this was the case. Of those with starting share capital, 73.1 per cent examine other companies' accounts compared with 32.3 per cent of other companies.

Share capital was not simply an alternative to other sources of funds: debentures were *more* likely to be important for companies using share capital companies at incorporation (25.9 per cent of those starting with share capital also used debentures, against 10.4 per cent using debentures overall). Presumably outsiders would not lend without the share capital commitment – a type of informal minimum share capital requirement. For the same reason, we might have expected banks to be looking for a reasonable share capital before lending, but our evidence from interviews was mixed on this. Examples of comments were, on the one hand, 'My partner is a financial wizard . . . his reputation helped with obtaining bank finance, so it was easy to start without personal capital' and on the other, from a company owner who started out with a £3,000 overdraft and an enterprise allowance, 'Banks are an influence on share capital. They like to see commitment. I have been encouraged to turn profit into share capital'. It might well be that in the current climate and with banks tightening up on lending, a greater emphasis is being placed now on share capital than was the case in the past. Our evidence supports this as an approach.

Clearly fitting well with theory was the fact that companies with share capital on incorporation were more likely to give limited liability to other shareholders (who might include shareholder-directors) as a reason for incorporating (44.4 per cent against 16.8 per cent overall) and also ease of share transfer (37.0 per cent against 17.6 per cent overall). They also perceived greater finance raising (40.7 per cent against 24 per cent overall) and accounting advantages (48.1 per cent against 35.2 per cent overall) of incorporation than other companies did. It is of interest that those with share capital were less concerned about confidentiality (37 per cent against 50.4 per cent overall thought lack of confidentiality a problem with corporate status).

Most of the differences cited above were significant on chi-squared or F-tests. Other differences were not statistically significant, but this may have been due to small sample size – they were still of interest as bearing out our hypothesis. Of those with share capital, 66.6 per cent thought the statutory audit a problem compared with 73.5 per cent of those without share capital. Of those with share capital, 14.8 per cent thought that the disadvantages of incorporation outweighed the advantages, as compared with 23.5 per cent of those without share capital.

It is arguable that many of these differences simply reflect the size of the business, and that the use of share capital at incorporation is a dependent variable influenced by business size. There is clearly some relationship but the indicators of size relate to the current position whilst the analysis of whether share capital was important relates to the date of incorporation; the temporal direction is such that the latter does not seem to be simply a surrogate for the former. The importance of share capital at the start could be an indicator of future size and growth potential. It seems that entrepreneurs planning to operate on a substantial scale are more likely either to put share capital into the company themselves or to seek it from outsiders *on incorporation* than are others. This provides a test which can be applied at incorporation before other indicators, such as turnover and profit, are available.

CONCLUSIONS

It has been shown that most companies are very small indeed (see Figure 7.1, p. 244 and Statistical Appendix, p. 270) and that companies incorporate for a variety of reasons: economic and legal, cultural and tax. It is a reasonable hypothesis that those incorporating for cultural and tax reasons are ultimately less likely to find the corporate form valuable than others and this is borne out by our findings on those dissatisfied with corporate form. Theory would suggest that incorporation to avoid being classified in a certain way would produce inefficiencies and dissatisfaction and this is also borne out by our results. For a sizeable minority of small companies, the 'cost' of operating in corporate form is likely to exceed the benefit. This results in inefficiencies for them and for those dealing with them. For example, third parties may require personal guarantees which they would not need from an unincorporated firm. Also, the ability of very small firms to incorporate may encourage them to enter trades without adequate capital. Even leaving aside the potential for abuse, this can have a negative effect for both those owning the business and those dealing with them. Measures to deal with corporate failures *after* they have occurred by imposing penalties on ex-directors or making them personally liable for the companies' debts may come too late, even if they can be effectively enforced. Our empirical findings therefore support the analysis of Halpern *et al.* (1980), referred to previously, that an unlimited liability regime would be the most efficient for small, tightly-held companies to avoid creating a moral hazard, transferring uncompensated risks to creditors, and inducing costly attempts to reduce these risks.

These findings contrast with the value placed on incorporation as a sign of serious business intentions both in small business literature and amongst the business community. The extent to which our respondents incorporated to obtain prestige, confirmed in face to face interviews, suggests that there is some 'reality' and perhaps some advantage to incorporating for this reason. However, since very small incorporated firms may be no more substantial than own account workers and since they are just as likely, if not more so, to fail to satisfy a creditor, there can be no long-term value in perpetuating the belief that they are automatically more substantial than unincorporated firms. Many of the more experienced business owners whom we interviewed were well aware of the difficulties of dealing with limited companies. The 'prestige' effect is therefore not straightforward and those incorporating purely for this reason, as well as those dealing with them, may well become disillusioned.

The small business literature which appears to suggest that incorporation *per se* may be an indicator of growth potential may also be in part an accurate reflection of the mistaken beliefs of others. Storey's findings (Storey 1992) that incorporated firms have a greater likelihood of obtaining bank lending at start up than others could simply reflect, as Storey points out, that the lending decisions of the banks during the 1980s may have been ill informed. The banks themselves may be or have been caught up in the culture which equates the limited liability company with entrepreneurship.[26] If banks have been relying on such indicators as legal status of the firm rather than the personal characteristics of the founder as a basis for lending decisions, this may have been unwise and have contributed to the current bad debt situation.

Storey also discovered that founders of (currently) *larger* new firms are more likely to opt for limited company status than unincorporated form *at start up*. This is surprising in that we would expect best advice often to be (mainly for tax reasons) to start in unincorporated form and incorporate at a later date.

Prior to incorporation, 40 per cent of our limited company respondents had traded in unincorporated form and there was evidence that those which did so were more satisfied with their legal form than those which went straight into corporate form.[27] It is clear from the VAT statistics that there is a link between incorporation and size (Figure 7.1) and Storey's results may simply be a reflection of this. This in itself could be related to a belief by some firms hoping to grow that they needed to be incorporated from the outset for prestige reasons as well as to enable them to raise finance. However, as discussed above, incorporation remains a crude indicator, since this figure will also include those very small companies with no intention of taking on employees or raising share capital. A breakdown of the corporate group into sub-categories could be of very great assistance in isolating indicators of growth and survival, but incorporation alone seems too wide. It is submitted that Storey's results do not show any inherent institutional efficiency of the corporate form which results in growth or survival; rather that certain firms will self-select the corporate form. These firms will include, but not be confined to, growth firms. As Hakim points out, although she finds that no growth firms are typically unincorporated and fast growth firms are typically incorporated, these characteristics are not mutually exclusive to a sufficient degree to allow us to make predictions about future expansion. Incorporation 'is not itself a guarantee or reliable predictor of future business development' (Hakim 1989).

Batstone's suggestion (1991) that companies should be targeted for assistance over unincorporated business is subject to the same criticism. Such a policy will include many non-growth firms whilst excluding those who start up in unincorporated form but still intend to grow, perhaps by incorporating at a later date. There is an additional objection to Batstone's proposal, that the availability of financial assistance to companies only, as he suggests, would distort the decision as to legal form and could result in incorporation which was not otherwise efficient. As shown by our data, the corporation is only likely to be the most efficient legal form where its attributes are actually required for what we have described as a 'legal or economic' reason.

Far from supporting further incentives to incorporate, our evidence suggests that incorporation of very small firms is not always appropriate. Some action could be taken to reduce the burdens on small companies; for example, as discussed above and elsewhere, the removal of the statutory audit requirement. However, this would do nothing to remove the misleading notion that incorporation can be equated with entrepreneurship and substance. This seems to be as much a matter of education of the business community as much as anything else. To ignore this could lead to bad decisions being taken on mistaken grounds.

A more fundamental question is whether it is desirable for the privilege of limited liability to be so easily accessible for very small firms on the face of it, only to be eroded by cost increasing personal guarantees and legislation to protect the public from abuse. This schizophrenic approach to limited liability – of the utmost value and importance on the one hand and a dangerous incitement to abuse on the other – leads to a system in which the foolish or ignorant could find themselves personally liable despite believing they are protected, whilst there is encouragement to set up business without personal commitment of capital, and scope for deliberate abuse remains. Our evidence is consistent with the hypothesis that by not requiring a minimum level of capitalization, UK company law encourages small business owners to start up on borrowed capital or with little capital behind them, whereas unincorporated business owners not only have unlimited liability but actually perceive themselves to be putting capital into their businesses; thus the different legal forms may have an important influence on attitudes to the amount of financial backing needed on start up. As shown, the importance of share capital at start up does seem to give some indication of the appropriateness of the corporate legal form. The banks should perhaps be more concerned with equity levels than they have been in the past. This could result in an informal minimum capital requirement.

A legal minimum capital requirement is a familiar notion in most European countries but has been rejected in common law countries as too easy to evade by returning cash to the promoters (Jenkins 1962) and because it is very difficult to set an appropriate level for all types of business. There would also be serious transitional difficulties on introducing such a requirement (DTI 1981). The most fundamental objection, however, underlying all these others, is that free accessibility to the corporate form is an idea embedded in our business culture and there would be a furore if an attempt was made to remove this privilege, despite the theoretically strong arguments for this and the fact that such a move could save business owners and third parties alike from business failures. A requirement of a substantial sum to be put into the company would drive home the seriousness of the endeavour even if the precise sum was hard to choose and no guarantee of a suitable capital level for all time. However, a move in this direction seems unlikely. If it were to come, it would come as a result of harmonization pressure from the EC, but there are no signs of this currently.

Unless we are prepared to make incorporation more difficult, by creating hurdles such as qualifying examinations (Hudson 1989) or minimum capital requirements, or both, then there is a limit to the penalties which should be imposed on those who simply do as encouraged by the rhetoric of the limited liability company. A cost may be described as a 'price' where there is a corresponding benefit but, where there is none, it is simply a penalty. This is the situation with the audit; hence the pressure for it to go. The Government is also investigating the possibility of simplifying the law affecting private companies to meet the same deregulatory pressure, although a new corporate form seems less likely than separate presentation of the law affecting private companies (DTI 1992a).

These moves, and other recent changes recognize the need to cater for very small companies but all run up against the problem of defining the companies to be targeted. Removal of the statutory audit requirement and disincorporation relief have both become over-complicated because of attempts to spread the net too wide, largely because of the difficulty of achieving a narrower, more appropriate definition. The attempt to simplify company law for all private companies may meet with the same problem. Further empirical and statistical work is necessary to enable policy and theory to be properly informed as to the wide variety of entities which make up the corporate sector. The category of 'small' private companies masks a great range of firms needing differentiated policies and analysis. Misperceptions about the limited liability company will only be removed completely by further sub-classification within this category.

STATISTICAL APPENDIX: THE NUMBER OF ACTIVE 'SMALL' COMPANIES

When commencing investigation of small limited companies, the authors' starting point was the Companies Act definition of 'small'.[28] There are no official statistics specifying the number of registered companies which satisfy the definition of 'small company' in the Companies Act 1985. Firms must qualify on two criteria out of turnover, employment, and balance sheet total. But it is not even possible to establish the position regarding each element of the tripartite definition because small companies (as defined in the Act) may opt to file modified accounts, in which turnover and number of employees need not be given. As a result, we have devised an estimate based on the figures which are available. In view of the fact that little or no analysis of employment or balance sheet structure is published for the smallest firms, we focus on turnover alone in the following estimates: although some companies with a turnover below £2 million will not qualify as small, because they exceed the small company limits on both the other two criteria (balance sheet total and number of employees), others will exceed £2m turnover, but will qualify on both employment and balance sheet total. These two factors will tend to counterbalance each other, so it can reasonably be assumed that taking the turnover criterion alone will yield a fair approximation of the total.

The total number of companies on the companies register in Great Britain in 1990–1 was 1,031,900 (DTI 1991); this omits Northern Ireland but the number of companies there is not significant for the purposes of this discussion. The best estimate of the number of companies reporting a turnover in excess of £2 million is in 'Size analyses of United Kingdom businesses' (CSO 1991), which provides information on the turnover of VAT-registered companies, broken down by legal form (Tables 3A–3D). However, the CSO only provides information on VAT-registered businesses: in consequence, much of the following discussion concerns the adjustments which are needed to reconcile the data on the VAT register with company information from other sources.

Since the CSO statistics do not show all companies below the VAT threshold (£25,400 in 1990–1, £36,600 in 1992–3), an important group for our purpose, the number of companies falling below the £2 million turnover limit can only be calculated by deducting the number of VAT-registered companies with a turnover above £2 million from the total number of companies on the companies register. The CSO gives a total of 55,290 VAT-registered companies reporting turnover in excess of £2 million in 1990, so we obtain a first estimate of some 977,000 (1,032,000 *minus* 55,000) companies below £2 million turnover in 1990. However, this estimate should be treated with considerable caution. As it has been obtained simply by subtracting the number of large VAT-registered companies from the total number of companies on the companies register, we need to establish that we have not omitted any large companies from our estimate, because such omissions will have the effect of inflating the number of small companies. In principle, errors could be caused by (i) VAT exemptions, where large companies do not appear on the VAT register at all; (ii) group VAT registrations, where several large companies register jointly for VAT; and (iii) divisional registrations, where one company registers each of its divisions separately for VAT.

VAT exemptions

Certain companies are exempt from VAT, either because they pursue an exempt activity, or because they are below a *de minimis* limit. There are eight exempt activities: firms trading entirely in insurance, postal services, education, health, property services, betting and funeral services may not register. A small number of totally VAT-exempt businesses over £2m will have been omitted from our estimate of the number of large companies because they are pursuing exempt activities; but because it is advantageous to register and obtain partial exemption if any of the business's activities are within the scope of VAT, total exemption on grounds of activity is likely to be uncommon among £2 million-plus businesses. The majority of firms which are totally exempt on grounds of activity are thus probably small. It should also be noted that *partly* exempt firms are classified on the basis of *taxable* turnover for VAT purposes; in consequence, such firms will sometimes be classified by the CSO as smaller than they really are. It is unlikely that this problem will greatly affect the total number of large companies.

There are a large number of businesses below the threshold for compulsory VAT registration: any company having a turnover below a *de minimis* limit (£25,400 in 1990–1, £36,600 in 1992–3) is not obliged to register. Since 1978–9, the VAT threshold has been maintained at a sufficiently high level to permit many of the smallest companies (and indeed many unincorporated small firms) to operate without registering for VAT. They may, however, opt to register: there were 243,242 voluntary registrations in 1990, of which 66,799

(27.5 per cent) were companies; many other very small companies will not have opted to register for VAT. Daly and McCann (1992) estimate that there may be one million small unincorporated firms not registered for VAT; and their estimate does not appear to include companies. By definition, no large companies could be exempt from VAT on turnover grounds.

Group registrations

Under VAT 'group registration' provisions, companies under common control may opt to be registered as a group, whereupon they become one 'legal unit' for VAT purposes. Where this option is taken, several companies under common control will appear on the VAT register as one registration. In consequence, the figure of 55,290 companies over £2m will underestimate the total number of 'large' companies to the extent of the number of subsidiary companies in group registrations which *individually* report more than £2m turnover. Thus, for example, if a group registration comprises four companies with a total turnover of £20m, consisting of two £8m companies, one £3m company and one £1m company, when we divide companies into 'large' and 'small' at our £2m cut-off point, the group will appear on the VAT register as one large company, but on the companies register as three large and one small. Then when we attempt to reconcile the two data sets, the companies register will show two more large and one more small company than the VAT register.

A further complication is that group VAT registration is an option, not an obligation: subsidiary companies may alternatively be registered as individual VAT legal units. So if, in our example, the group had not opted for group registration, it would appear identically on the VAT register and on the companies register and there would be no reconciliation problem! To quantify the extent of this discrepancy, we need to know:

1 The total number of group registrations (item a).
2 The total number of subsidiary companies (item b1); whether they have a turnover in excess of £2m (item b2.1) or not (item b2.2); and how they are divided between group registrations (item b3.1) and separate registrations (item b3.2).

There were 24,000 group accounts registered at Companies House in 1990–1 (DTI 1992). Daly (1990) states that there are about 20,000 VAT group registrations (item a). (Although there are differences in the relevant VAT and Companies Act definitions, there is a broad equivalence sufficient for our purposes here.) Page (1984) found in a sample of 1,000 companies drawn from the *companies* register that 15 per cent were subsidiaries. Using Page's findings, we can estimate that there were approximately 155,000 (15 per cent of 1,031,900) subsidiary companies altogether (item b1). Ganguly (1985) reports that less than 4 per cent of the individually-registered VAT legal units are subsidiaries. Taking Ganguly's findings, we can estimate that approximately 70,000 (3.9 per cent of 1,765,178, the total number of VAT registrations) subsidiaries were individually registered for VAT (item b3.2), leaving 85,000 companies (155,000 − 70,000) which were presumably included in VAT groups (item b3.1). This estimate yields a plausible average of three-and-a-half companies per VAT group registration. However, we still have no data on the size distribution of subsidiaries (items b2.1 and b2.2). If we assume that *all* the 85,000 subsidiary companies about which we have no turnover information report turnover in excess of £2 million (i.e. that b2.1 = 85,000 and b2.2 = 0, a highly improbable assumption), we obtain an upper limit of 140,000 (55,000 + 85,000) 'large' companies. This unrealistically high estimate of large subsidiary companies should also be more than adequate to compensate for any large exempt companies omitted from the VAT register.

Divisional registrations

Daly (1990) states that there are 500 divisional registrations. This is the only circumstance in which the VAT register counts a company more than once. Thus in principle it would be possible for a large company to appear on the VAT register as several small companies. In practice, it is likely that this option is mainly taken by very large companies, and that most such divisions would individually exceed £2m turnover. In view of the fact that the number of divisional registrations is small, this factor is unlikely to affect the overall estimates to any great extent.

Estimates of the numbers of small and large companies

Bearing in mind that we have, if anything, overestimated the number of large companies, we can make a minimum estimate of some 892,000 (1,032,000 *minus* 140,000) small companies below £2 million turnover. In other words, in 1990, the majority of UK companies (86.4 per cent) probably satisfied the small companies criteria. This

274 Judith Freedman and Michael Godwin

compares reasonably well with the Government's 1985 estimate that 90 per cent of companies would be classified as 'small' under the thresholds it then proposed (DTI 1985: para. 8.5).

In connection with the debate on removing burdens on small companies, others have produced estimates of around 400,000–500,000 'small' companies within this definition (ICAEW 1992; Carsberg *et al.* 1985: 15). But if such estimates were accurate, we would be left with an unbridgeable gap of 392,000 between the known total of companies on the register (1,032,000) and the total of small plus large companies (500,000 + 140,000 = 640,000). These estimates are apparently based on the view that at least 90 per cent of active independent companies are small, but attempt to delete non-active companies and those which are part of a group. Whilst it is useful to delete the latter because we shall not want to target them as small companies for deregulation or assistance purposes, many so-called non-active companies (for example, flat management companies which do little more than hold the freehold of a block of flats) are still subject to Companies Act requirements, for example to have their accounts audited and the debates on these issues should not exclude such companies. Carsberg *et al*'s estimate may be based on the survey by Page (Page 1984), in which he found 41 per cent of a sample of 1,000 companies to be active and independent. However, this percentage excluded new companies which had not filed accounts (23 per cent) and various other groups for which the statutory audit issue is highly relevant. As noted above, he found 15 per cent of his sample to be subsidiaries. We suggest, therefore, that the number of companies which are 'small' under the Companies Act 1985 is greater than that often cited. The exclusion of new companies and others which are not trading but are active in some other way misleads as to the proportion of companies which come within this definition and leads to an inaccurate impression of the balance between companies of different sizes in existence.

Estimating the number of companies subject to annual audit

To estimate the number of small companies currently subject to annual audit, we start from our estimate of 892,000 such companies. From this figure we must deduct 78,100 companies which were dormant, and so were not required to produce audited accounts, in 1990–1 (DTI 1992). We must also exclude subsidiaries, using Page's finding that 15 per cent of his sample were subsidiaries. However, some of these will be large companies: the assumption which we have made is that 85,000 of an estimated 155,000 subsidiaries were large (over-£2m) companies, and 70,000 were small; this would leave 744,000 small, independent, non-dormant companies on the Register, all requiring a statutory audit even though they might not all be very active. However, it is highly probable that less than 85,000 subsidiaries were large: to the extent that this was the case, the number of small independent companies would be reduced.

Table 7.A1 The number of independent small and large companies in 1990

Category	Number	Source
Total number of companies	**1,031,900**	DTI
Companies over £2m turnover	**55,290**	CSO PA1003
Subsidiary companies in VAT group registrations; assume further that these are all over £2m turnover	85,000	Page finds subsidiaries constitute 15% of the companies register; Ganguly finds that fewer than 4% of the individually registered VAT units are subsidiaries
Total over £2m turnover	140,000	
Total £2m or below	892,000	
Subsidiary companies registered individually for VAT; assume further that these are all £2m or below	70,000	Ganguly finds that fewer than 4% of the individually registered VAT units are subsidiaries
Dormant companies	**78,100**	DTI
Over-£2m companies subject to audit	**55,290** plus any large exempt companies	CSO
Up to £2m companies subject to audit	744,000	
Total companies subject to audit	799,000	

Notes: 'Small' is up to £2m turnover. Estimates are in light type, official statistics in **bold** type.

METHODOLOGICAL APPENDIX

Information was obtained from 429 firms in four geographical areas using a combination of mail questionnaires and telephone interviewing. The sample included equal numbers of incorporated and unincorporated firms. Information from the survey was backed up by company searches and twenty-four face to face interviews with selected firms. Interviews were also conducted with twelve firms of auditors used by our survey respondents (although individual companies were not discussed with them). Professional firms were excluded since these are frequently regulated as regards legal form. One hundred and twenty-six usable completed questionnaires were received from limited companies (one of which was atypical and was therefore subsequently excluded from analysis) and 146 from unincorporated firms. Of the unincorporated firms, eighty-two were sole traders and sixty-four partnerships. Of the telephone respondents, fifty were sole traders, thirty-seven partnerships and sixty-five had limited companies. Five telephone respondents stated that they operated in both forms. The postal questionnaires gave the impression of being very conscientiously completed and, where answers were checked by reference to a company search or personal interview, this impression was confirmed.

The principal method was a pre-piloted mail questionnaire survey of 880 small unincorporated firms and 880 limited companies, supplemented by telephone and face-to-face interviews with some respondents and auditors. Findings were corroborated by other information, and responses were checked for credibility; company searches were carried out on approximately 300 of the sample companies, including all respondents and a random sample of non respondents.

Two separate questionnaires were designed; one for limited companies and one for unincorporated firms. The alternative would have been to send one very long questionnaire to all sample members with almost half the questions being inapplicable to any individual firm. However, efforts were made to include parallel questions in the two questionnaires. The focus of the questionnaires was the reason for the owners' decision as to legal form and the level of satisfaction with this form, but other questions addressed background information of relevance, such as advice taken, size and profitability of the firm and reaction to recent deregulation measures. Printed questionnaires were posted to two areas in September 1990. The posting to the second two areas took place in November 1990, for reasons connected with funding and resources. The final return was received in April 1991. In the light of pilot experience, it was decided that the first reminder should be by telephone: if a written response appeared unlikely, an attempt was made to ask the principal questions over the telephone. Telephoning was conducted by a team of students. Telephone interviews were analysed to discover whether those who had not completed the questionnaire differed from those who had; they also facilitated identification of definite refusals and of those no longer at the trading address. Three weeks after the telephone calls, a final postal reminder was sent out incorporating a further copy of the questionnaire. Tracking down telephone numbers of limited companies proved difficult: many were not trading under their registered names; and those using accommodation addresses seemed especially difficult to reach. The company responses included some which were originally in the unincorporated firm sample, were discovered to be incorporated, and agreed to complete a limited company questionnaire. A few responses were received from firms outside the sample but having the same owners as the sampled firm; some were larger than our turnover ceiling, but no responses were excluded from analysis on size grounds.

Although the questionnaire response was low, as in the pilot, the combined response rate for firms which completed a questionnaire or gave a usable telephone interview is a respectable 29 per cent (making adjustments based on the numbers of companies found to be out of frame through telephone and company search work), reasonable for small business research (Sandford, Godwin and Hardwick 1981). The pilot responses, plus a questionnaire from a company which volunteered to take part, are not included in the response rate but are included in the analysis.

NEW DEVELOPMENTS

Since this chapter was written, there have been a number of relevant changes and developments. It was announced in the November 1993 Budget speech that the mandatory statutory audit requirement will be abolished for companies with a turnover of less than £90,000. Companies with a turnover of between £90,000 and £350,000 a year will need only an independent accountant's report on whether the company's accounts correctly reflect its books. The full details of the scheme have yet to be published. There has been some criticism of these thresholds as arbitrary, or too low. This illustrates the author's point that the Companies Act definition of a small company is too wide as a basis for special reliefs. However, the formulation of alternatives is contentious and problematic.

The 1994 Finance Bill contains detailed new provisions of self assessment for sole traders and partnerships for the purpose of income tax which will result in their moving over to a current year basis. The change has immediate effect for businesses starting after

5 April 1994 and existing businesses will move to the new basis for the tax year 1997–8.

The 1994 Finance Bill also provides that for the financial year 1994 the thresholds for application of the small companies rate of tax (referred to at page 248 and footnote 15 above) will be £300,000 for full relief with tapering relief for profits up to £1,500,000.

On the EC front, the European Commission is known to be considering the pros and cons of extending the Second Directive on Company Law (which includes provisions about minimum capital) to private companies. However, this is a long way from being a proposal at present. There are also plans to *increase* the permitted small company thresholds in companies legislation. Clearly any UK proposals for reform must be considered within this EC context.

NOTES

* The authors would like to express thanks to the Department of Trade and Industry which, for this particular project, supplemented their funding.
1 During the course of the authors' project, this definition required that a 'small company' should not exceed two of the following thresholds: balance sheet total not more the £975,000; turnover not more than £2 million; number of employees not more than 50 (s.247 Companies Act 1985 as amended). The first two thresholds were increased to £1.4 million balance sheet total and £2.8 million turnover in November 1992 by the *Accounts of Small and Medium-Sized Enterprises Regulation 1992* ((1992) S.I. No. 2452).
2 King (1977) finds that the decision to incorporate is governed by four factors: (i) limited liability, (ii) separation of ownership from control facilitating free transfer of shares (iii) pension provisions and (iv) taxation. Limited liability is recognized as being partly illusory and is seen primarily as a means of reducing the cost of monitoring management. In his work on new form formation, Robson uses these factors, admittedly adapting the first, to investigate the relationship between the stock of self-employment and the level of company incorporations in Great Britain. However, the factors set out by King are inadequate to explain why *small* firms choose to incorporate. This must cast some doubts on Robson's analysis. The very posing of the question of the relationship between self-employment and company incorporations, and the equation of incorporation with 'new firm births' in his article, denies the importance of unincorporated firms and suggests a qualitative difference between small companies and unincorporated firms which reveals an incomplete view of the corporate form.
3 The contractual analysis of the company is, in any event, now the subject of considerable debate and criticism in the legal literature – see Bratton (1989) and generally the debate in vol. 89 *Columbia Law Review* (1989).
4 The difficulty, of course, is to determine what is legal policy if it is not based on efficiency. Teubner argues that 'justice has to control economic efficiency' and 'justice can be reformulated as a legal balancing of the internal consistency of law against different rationalities in society'.

5 For the decided cases on this see Gower 1992, chapter 6.
6 There has been considerable discussion of the need for a new legal form specially tailored to small firms in the UK. The history of the debate is described in Freedman and Godwin (1991). The authors have concluded elsewhere (Freedman and Godwin 1992) that there would be severe difficulties in gaining acceptance for a new legal form without limited liability if introduced into the UK in the current climate, all else being equal. The need for enhancement of unincorporated status and/or modification of the corporate form to provide a suitable vehicle for such firms will be considered further in a future article and is not therefore pursued here.
7 Recent work goes further and argues that there may be no reason to prefer limited liability over a regime of unlimited pro rata shareholder liability for corporate torts for either closely-held or publicly traded corporations, although this argument is not extended to contractual creditors (Hansmann and Kraakmann 1991).
8 Removed by *The Companies (Single Member Private Limited Companies) Regulations 1992* (S.I. 1992 No. 1699).
9 At the time of the survey. This threshold would clearly need to be increased for inflation.
10 As explained, our sample selection process excluded some companies with a turnover in excess of £1 million. However, there was a response bias towards our top turnover bands (Freedman and Godwin 1991). It is not claimed here that the size characteristics of our respondents precisely reflect that in the population at large. What is shown is the existence of a group of companies with characteristics as described and it is this type of company which we explore further here.
11 Of those providing information, 81 per cent mentioned banks (including 7 per cent who also mentioned others). Nine per cent mentioned others (for example, suppliers and landlords) only. The remainder did not specify the nature of the guarantee.
12 Although see R. Maas (1990) at para. 2.23, 'Accordingly in most cases the businessman will not achieve the protection of limited liability as against his bank by using a company. He will nevertheless generally reduce his exposure to the bank in so far as it will normally take a debenture over the company's assets so that in the event of a failure of the business whatever assets remain (after paying preferential creditors) will go to the bank rather than the general body of creditors, so reducing the entrepreneur's exposure under his guarantee'.
13 See Gower (1992) at p. 46 'The mystic word "Limited" was intended to act as a red flag warning the public of the perils which they faced if they had dealings with the dangerous new invention'.
14 See also Bolton Report, chapter 2. Note also Hakim (1989) –unincorporated firms have fewer staff and are less likely to have separate business premises.
15 There is a tapering relief for companies with profits up to £1,250,000. The Income and Corporation Taxes Act 1988 definition of small company is thus quite different from the Companies Act definition.
16 Note that the 1993 Finance Act reduces the tax credit on dividends to 20 per cent with the consequence that higher rate taxpayers have increased liability on dividends and the system is less fully integrated than previously. This tips the tax balance further in favour of unincorporated

firms. The change also illustrates the dangers of choosing legal form by reference to a tax system, the effects of which can change very rapidly.

17 See note above for most recent changes.

18 It is possible that dissatisfied firms were more likely to respond to the questionnaire than others. A follow-up telephone survey was conducted to test for bias: 157 non-respondents were contacted by telephone, and 22 questionnaire respondents were also contacted to validate their responses. The 68 telephone calls to non-respondent companies discovered that only 7 per cent of these were dissatisfied with their legal form, one-third of the questionnaire percentage. However, the 89 telephone calls to non-respondent unincorporated firms discovered no dissatisfaction at all. So even if the questionnaire response is disproportionately high on this issue, the *relative* findings – that small companies are less satisfied than unincorporated firms – were borne out by the telephone survey. Assuming the greatest response bias, that is that the telephone response represents 80 per cent of the limited population, the percentage of dissatisfied companies is still 10.2 per cent.

19 *Fall v. Hitchen* [1973] STC 66.

20 Truman ((1990), chapter 4. 'Starting a Business' states 'Regrettably, there are certain industries which are only prepared to utilise the services of a limited company; the most notable being the computer programming industry'.

21 One of our telephone sample informed us that her son had decided not to start up on his own because he could not get work without forming a company, action which he was reluctant to take.

22 Respondents were not asked to state amounts of finance contributed in each way as our interest was in their perceptions and the importance of the source to the particular business (and because we did not wish to ask an undue number of sensitive questions which might have detracted from the response to our central questions).

23 Should the business fail they may reclaim their loan, which may even be secured; *Salomon v. Salomon & Co Ltd.* [1897] AC 22, although now see s.245 Insolvency Act 1986 which provides for the avoidance of floating charges in favour of persons connected with the company in certain circumstances.

24 The results from surveys conducted on behalf of the Bolton Committee showed that unincorporated businesses had a lower proportion of debt in relation to total assets than did companies; in particular, the unincorporated firms in this sample tended to borrow less from banks than did companies. The Committee's explanation was that 'Apart from the effect of size, the inability of a lender to obtain a floating charge from an unincorporated borrower would lead us to expect lower borrowing ratios for the unincorporated firm'. It may be that in the case of our respondents, who were smaller than the Bolton respondents, there was little over which the bank could take a floating charge (generally stock and debts). This lack of corporate assets might mean that the lender would take a fixed charge over the individual owner's home in any event in which case incorporation makes no difference to the lender.

25 It is possible that a larger sample might have revealed sectoral differences but no relevant statistically significant differences appeared from our data.

26 Storey himself suggests, based on Cressy (1992), that most wholly new firms obtaining loan/overdraft facilities from the largest UK banks were not required to provide security because the sums involved were so small. Again, this is surprising in view of our finding that over half our company respondents provided personal guarantees, but this does not support the Bolton Committee theory, outlined above, that lending to companies is preferred by banks because of the availability of the floating charge.

27 Of our company respondents overall, 58.4 per cent were incorporated from start up, whereas 70 per cent of our dissatisfied companies started out as incorporated firms.

28 See note 1 above.

REFERENCES

Batstone, S. (1991) 'The significance of legal form for growth in business services: evidence from the UK and Norway', Paper presented to *21st European Small Business Seminar*.

Binks, M. (1991) 'Small businesses and their banks in the year 2000' in Curran, J. and Blackburn, R.A. (eds) *Paths of Enterprise – The Future of the Small Business*, Routledge, London.

Boggan, S. (1993) 'Inside story', *Independent on Sunday*, 7 March, p. 16.

Boggan, S. (1993) 'Review of accounts reveals saving of £810', the *Independent*, 9 March.

Bolton Report (1971) *Report of the Committee of Inquiry on Small Firms*, Cmd. 4811, HMSO, London.

Bratton, W.W.Jnr (1989) 'The "nexus of contracts" corporation: a critical appraisal' *Cornell Law Review* 74, 407.

Carsberg, B.V., Page, M.J., Sindall, A.J. and Waring, I.D. (1985) *Small Company Financial Reporting*, Prentice Hall, ICAEW.

Central Statistical Office (1991) *Business Monitor PA1003*, CSO, London.

Chesterman, M. (1982) *Small Businesses*, Sweet & Maxwell, London (2nd edition).

Coase, R.H. (1937) 'The nature of the firm', *Economica 4*, November, p. 386.

Collins, H. (1990) 'Independent contractors and the challenge of vertical disintegration to employment protection laws', *Oxford Journal of Legal Studies*, 10, 3, 53.

Cork Gully (1993) *Annual Review*, Cork Gully, London.

Cressy, R.C. (1992) *Business and Proprietor Characteristics, Complementary Financial Sources and Bank Lending: The Case of UK Business Starts*, University of Warwick SME Centre, Warwick.

Daly, M. (1990) 'The 1980s – a decade of growth in enterprise' *Employment Gazette*, November, pp. 553–65.

Daly, M. and McCann, A. (1992) 'How many small firms?' *Employment Gazette*, February, pp. 47–51.

DTI (1981) *A New Form of Incorporation for Small Firms*, Cmnd 8171, HMSO, London.

DTI (1985) *Building Businesses . . . Not Barriers*, Cmnd 9794, HMSO, London.

DTI/Inland Revenue (1987) *Disincorporation*, DTI/Inland Revenue, London.

DTI (1988) 'Consultative Document on Twelfth Directive', DTI, London.
DTI (1991) *Companies in 1990–91*, DTI, London.
DTI (1992) *Companies in 1991–92*, DTI, London.
DTI (1992a) 'Press Release on Improving Company Law', 18 November.
Easterbrook, F.H. and Fischel, D. (1985) 'Limited liability and the corporation', *The University of Chicago Law Review*, 52, 89.
Easterbrook, F.H. and Fischel, D. (1986) 'Close corporations and agency costs', *Stanford Law Review*, 38, 265, 271.
EC (1989) *Twelfth Council Company Law Directive* (89/667/EEC).
Fama, E.F. and Jensen, M.C. (1983) 'Agency problems and residual claims', *Journal of Law and Economics*, XXVI (June).
Farrar, J., Furey, N. and Hannigan, B. (1991) *Company Law*, Butterworths, London, 3rd edition.
Finch, V. (1990) 'Directors' disqualification: a plea for competence', 53 *Modern Law Review* 385.
Freedman, J. and Godwin, M. (1991) *Legal Form, Tax and the Micro Business*, Institute of Advanced Legal Studies Working Paper.
Freedman, J. and Godwin, M. (1992) 'Legal form, tax and the micro business' in Caley, K., Chell, E., Chittenden, F. and Mason, L. (eds) *Small Enterprise Development*, Paul Chapman Publishing, London.
Freedman, J. and Godwin, M. (1993) 'The statutory audit and the micro company – an empirical investigation' *Journal of Business Law*, March, 105.
Gable, S. (1989) 'One Man Band', *Taxation*, 26 October.
Ganguly, P. (1985) 'How small is a small firm?' in Bannock, G. (ed.) *UK Small Business Statistics*, Harper & Row on behalf of the Small Business Research Trust.
Gower, L.C.B. (1992) *Principles of Modern Company Law*, Sweet & Maxwell, London.
Gray, C. (1992) 'Growth-orientation and the small firm' in Caley K., Chell, E., Chittenden, F. and Mason, L. (eds) *Small Enterprise Development,* Paul Chapman Publishing, London.
Hakim, C. (1988) 'Self-employment in Britain: recent trends and current issues' *Work, Employment and Society*, 2, 4, 421.
Hakim, C. (1989) 'Identifying fast growth small firms', *Employment Gazette*, 29, January.
Halpern, P., Trebilcock, M. and Turnbull, S. (1980) 'An economic analysis of limited liability in corporation law', 30 *University of Toronto Law Journal* 117.
Hansmann, H. and Kraakman, R. (1991) 'Toward unlimited liability for corporate torts', *Yale Law Journal* 100, 7.
Hudson, J. (1989) 'The limited liability company: success, failure and future', *Royal Bank of Scotland Revue*, no. 161, March.
ICAEW (1992) *The Statutory Audit of Small Companies – The Case for Reform*, FRAG 21/92.
ICAEW (1993) 'The statutory audit of small companies' (report on responses to FRAG 21/92), FRAG 4/93.
Inland Revenue (1991) *A Simpler System for Taxing the Self Employed,*
Inland Revenue (1992) *A Simpler System for Assessing Personal Tax.*
Institute of Directors (IOD) (1986) *Deregulation for Small Private Companies.*
Jenkins (1962) *Committee on Company Law*, Cmnd 1749, HMSO, London.

Jensen, M. and Meckling, W. (1976) 'Theories of the firm: managerial behaviour, agency costs and ownership structure', *Journal of Financial Economics*, 3, 305.
King, M. (1977) *Public Policy and the Corporations*, Chapman & Hall, London.
Maas, R. (1990) *Tax Planning for the Smaller Business*, Butterworths, London.
McGregor, A. and Sproull, A. (1992) 'Employers and the flexible workforce', *Employment Gazette*, 225, May.
McNulty, P.J. (1984) 'On the nature and theory of economic organization: the role of the firm reconsidered', *History of Political Economy* 16, 2, 233–53.
Manne, H.G. (1967) 'Our two corporation systems: law and economics', *Virginia Law Review*, 53, 259.
Page, M. (1984) 'Corporate financial reporting and the small independent company', *Accounting and Business Research* (Summer).
Paillusseau, J. (1991) 'The nature of the company' in Drury, R. and Xuereb, P. (eds) *European Company Laws*, Dartmouth, Aldershot.
Posner, R. (1986) *Economic Analysis of Law*, Little, Brown and Company, Boston and Toronto, 3rd edition.
Rainbird, H. (1991) 'The self employed: small entrepreneurs or disguised wage labourers?' in Pollert (ed.) *Farewell to Flexibility?*, Basil Blackwell, Oxford,
Robson, M.T. (1991) 'Self Employment and the New Firm Formation', *Scottish Journal of Political Economy* November, 38, 4.
Sandford, C. Godwin, M. and Hardwick, P. (1981) *Costs and Benefits of VAT*, Heinemann, London.
Sandford, C., Godwin, M. and Hardwick, P. (1990) *Administrative and Compliance Costs of Taxation*, Fiscal Publications.
Storey, D.J. and Johnson, S. (1987) *Job Generation and Labour Market Change*, Macmillan Press, London.
Storey, D.J. (1992) 'New firm growth and bank financing', SME Centre University of Warwick, May.
Teubner, G. (forthcoming 1994) 'Piercing the contractual veil? – The social responsibility of contractual networks' in Wilhelmsson *Critical Contract Law*.
Truman, M. (1990) *The Practitioners' Tax Manual*, Gee and Co., London.
Wood, D. and Smith, P. (1989) *Employers' Labour Use Strategies – First Report on the 1987 Survey*, Research Paper no. 63, Department of Employment, London.

8 Acquisition activity in the small business sector*

Andy Cosh and Alan Hughes

INTRODUCTION

It is well known that mergers and acquisitions play a central role in the growth and development of businesses. The process is extremely well documented for large firms, especially those with quotations on stock exchanges. There is a large literature analysing their scale, their variation over time, their determinants and their effects (Auerbach 1988, Hughes 1993). Within this literature the role of smaller businesses has been relatively neglected, due in part to both lack of data availability and methodological problems. A few studies have noted that large firms may buy into portfolios of smaller firms. In addition the rate at which such firms are bought and then sold off and their subsequent performance has been analysed in relation to the US conglomerate merger boom of the 1970s (Ravenscraft and Scherer 1987). This is one of the few systematic studies of mergers involving smaller firms since the work of the Bolton Committee in the UK in the late 1960s. In their US study Ravenscraft and Scherer found that small acquired companies were typically outperforming their industry averages prior to takeover. They found that sellers initiated the takeover for a mixture of three motives: solving problems of management succession; converting personal stock to more liquid assets; and gaining access to better financial resources. They were only rarely driven by financial failure. Taken as a group, acquired firms suffered post-merger falls in relative profitability due either to management control losses, or to exploitation of the acquired firms as a cash cow; the worst performers were then sold off. Acquisitions not sold off subsequently tended unsurprisingly to be better post-merger performers, but not to the extent of recording gains in profitability relative to their non-merging control groups. Only acquisition between equal-sized firms seemed to offer systematic gains in this respect, possibly because top management in both parties to the merger were equally committed to success (their focus on large conglomerate acquirers meant that Ravenscraft and Scherer had relatively little to say about smaller firms as acquirers).

Evidence for the UK is older and less systematic (For a review see Hughes 1991). The UK studies produced around the time of the Bolton Report echo the seller motivations discussed above. Thus in a study of smaller firms in the East Midlands hosiery and knitwear industries (Boswell 1972) the desire for realized capital gains and the solution to problems of management succession were the most frequently cited reasons for being open-minded about the prospect of selling out. However, in actual cases of sell out, other studies suggest that these reasons are less significant than financial problems and the need to access funds (Merrett Cyriax Associates 1971). This latter finding was echoed in a recent study of barriers to growth in small firms, albeit on the basis of a small number of case studies (ACOST 1990). Analysis of post merger effects for small firms is even more sparse for the UK and we have no profitability, or other accounts-based studies, stretching beyond the larger quoted population.[1]

These gaps in our knowledge are of significance not only because of the scale of involvement of smaller firms in acquisition activity, which will be demonstrated in this chapter, but also because of the increased policy interest in this area. This reflects a perception that high rates of small firm acquisition 'deaths' may contribute to the relative lack of medium-sized, independent companies in the UK in comparison with other European countries (ACOST 1990, Barber, Metcalfe and Porteous 1989), with potential implications for overall economic performance. There has thus been some concern that high-tech small firms fail to prosper in the UK and do not emerge as international players because financial, management and other constraints drive them to sell out to larger (often overseas) firms, with potentially deleterious effects on their innovation performance (ACOST 1990, Garnsey and Roberts 1990 and Garnsey, Alford and Roberts 1990). Nearly all of this UK literature is based on very small samples of firms, indeed often on case studies of single firms studied intensively over a period of time. Furthermore we do not have many attempts to document systematically the size distribution of merger activity in the UK, the proportionate significance of smaller firms within it, the motivations of buyers and sellers, the effects of small company acquisitions and their relative success compared with acquisitions involving solely large companies.[2] This chapter attempts to fill these gaps.

The next section critically assesses measures of the scale of acquisition activity amongst small firms. We then turn to the motivation and

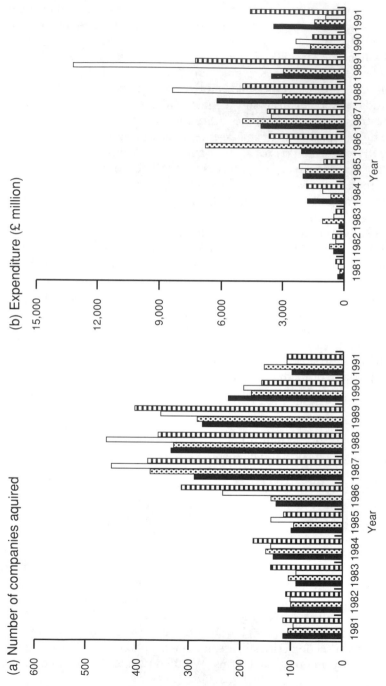

Figure 8.1 UK acquisition and mergers of industrial and commercial companies: quarterly data

attitudes of small firms, both as acquirers and as targets, making extensive use of a recent national survey of over 2,000 SMEs (SBRC 1992) which *inter alia* specifically addressed these issues. We then assess the outcome of the acquisition process by first identifying the characteristics of acquiring and acquired firms and then by examining the effect on the acquired company. This additionally draws upon a specially constructed panel database of several thousand companies in the period 1977–82, which allows a comparison of large and small buyers and sellers relative to their industries. This is supplemented by a detailed examination of qualitative data obtained from micro-fiches of a sample of company reports. Finally the chapter ends with a summary of our findings in comparison with acquisition studies of large firms and a discussion of the future direction of research in this area.

SCALE OF ACTIVITY

The pattern of acquisition expenditure and the number of companies acquired by UK industrial and commercial companies is shown in Table 8.1 for the period 1969– 91. This series is based on official data which for smaller companies is collected primarily from information in the financial press. The volatility of acquisition behaviour and the burst in activity between 1986 and 1989 is clearly visible. This is portrayed in Figures 8.1a and 8.1b which show the quarterly pattern in both expenditure and numbers acquired over the past decade.

In order to assess whether this pattern is also observable amongst small firms we need to separate the acquisitions by the size of target firm. In Table 8.2 acquisitions have been classified according to the total consideration paid for the acquired firm. The years have been

Table 8.1 Acquisition and mergers by industrial and commercial companies within the UK 1969–91

Annual Averages	Number Acquired	Expenditure (£m)	Expenditure in 1962 stock market prices (£m)
1969–73	988	1,388	777
1974–81	459	947	426
1982–5	488	4,278	777
1986–9	1,301	20,500	1,930
1990	778	8,252	688
1991	498	10,574	860

Sources: Business Monitor MQ7, *Business Bulletin: Acquisitions and Mergers within the UK*, Economic Trends Annual Supplement.

Table 8.2 Size distribution of acquisitions and mergers of independent companies by industrial and commercial companies within UK 1985–91

Consideration Paid	Number	%	Expenditure	%
1985				
⩾ £25m	31	9.1	5,288	84.0
£10–25m	30	8.8	500	7.9
£1–10m	123	36.2	452	7.2
⩽ £1m	156	45.9	57	0.9
1988				
⩾ £100m	29	2.6	12,047	70.0
£10–100m	111	9.9	3,379	19.6
£1–10m	480	42.7	1,585	9.2
⩽ £1m	503	44.8	199	1.2
1991				
⩾ £100m	10	3.5	6,069	79.1
£10–100m	36	12.6	1,094	14.2
£1–10m	122	42.8	481	6.3
⩽ £1m	117	41.1	31	0.4

Sources: see notes to Table 8.1.

chosen to represent both acquisition-slack years, 1985 and 1991, and the acquisition-intensive year, 1988. What emerges from this table is that small acquisitions appear to follow the same pattern which has been identified for larger acquisitions. Thus when considering acquisitions for which the consideration paid was less than £1m we find a marked rise between 1985 and 1988 in the scale of activity, in terms of both the numbers acquired and the amounts spent, and a subsequent decline to 1991. This finding suggests that factors which are common to both large and small firms, such as the level of demand and financial conditions, are likely to explain the pattern of activity we observe. The other important finding is the scale of merger activity amongst small companies – over 40 per cent of acquisitions are for a consideration of £1m or less. This point is explored further below using data from another source.

An alternative source of information about acquisition activity comes from analysis of the number of businesses on the VAT register. It is important to note that the register includes sole proprietorships and partnerships as well as companies. Such an analysis is presented in Table 8.3 which allows us to site acquisition deaths amongst other forms of business deregistrations and to provide some broad industrial analysis. The figures again reveal a greater intensity of M & A deregistrations in 1986–9 than either

Table 8.3 VAT register – deregistration of business 1982–90

	Average annual rate of de-registration %				
	Production	*Construction*	*Distribution*	*Other*	*Total*
1982–5					
M & A deregistrations	1.4	0.5	4.0	22.1	2.3
Failed company deregistrations	8.4	7.4	7.3	5.9	6.8
Other deregistrations	1.6	2.7	1.7	1.9	1.9
1986–9					
M & A deregistrations	1.7	0.6	4.1	2.4	2.5
Failed company deregistrations	7.6	7.0	7.2	6.0	6.6
Other deregistration	2.2	3.3	2.0	2.1	2.3
1990					
M & A deregistrations	1.5	0.5	2.8	1.8	1.8
Failed company deregistrations	6.8	6.6	7.0	6.2	6.5
Other deregistrations	2.0	3.2	1.9	2.2	2.3
Number of M & A deregistrations					
1982	1,817	957	15,995	14,032	32,401
1983	1,829	1,012	15,490	13,963	32,294
1984	1,864	1,055	14,170	13,874	30,963
1985	1,852	1,129	14,363	14,456	31,800
1986	2,119	1,180	14,870	15,462	33,631
1987	2,353	1,297	15,321	17,584	36,555
1988	2,606	1,585	16,662	19,304	40,157
1989	2,697	1,510	14,502	17,808	36,517
1990	2,352	1,298	10,909	15,380	29,939

Note: Rate of deregistration is the number deregistering as a percentage of the opening stock of registrations

before or after that period, but the effect is not as marked as found from other sources. The scale of acquisition activity shown is massive. It is clear that the vast majority of acquisitions go unreported in the official statistics relating to industrial and commercial companies. It is worth noting that failed business deregistrations do not reveal the same pattern as acquisitions, but show some tendency to decline from 1982 to 1990. This cause of deregistration is quantitatively much more important than M & A deregistrations. The greater intensity of acquisition activity in the distribution sector can be contrasted with the low levels of activity (only one in two hundred per year on average) in the construction sector.

A further source of information is the DTI stratified sample of UK registered companies which provides a more reliable comparison of the scale of acquisition activity amongst large and small firms. Whilst the Business Monitor series provides data on acquisition numbers and total consideration paid, it provides no denominators to allow the series to be normalized by the numbers of companies in various size classes. Equally, whilst the VAT register has a wider coverage of businesses, and deaths can be normalized by the number of businesses on the register, it does not provide a size analysis of deaths. The DTI sample by comparison allows normalization by size classes. This was done for the period 1977–82 by examining company records lodged at Companies House. The information concerning the number of employees for firms within this sample was patchy and so the division between small and larger companies has been chosen as total assets in 1976 of £1m. The findings are presented in Table 8.4 and show a similar ratio of death by failure to acquisition death as that found from the analysis of the VAT register. The survival rate of larger companies (78.5 per cent) is noticeably greater than that for small companies (68.7 per cent). Further analysis of overall survival rates within this sample reveals that the lower death rate amongst large firms occurs within the upper quartile of firm size and that there is little difference between small and medium companies in overall death rate. On the other hand smaller companies had in this period a much lower likelihood of being acquired (3.5 per cent) than larger companies (12.0 per cent). Therefore for this period at least we find small companies much more likely to fail, but less likely to be acquired than the rest of the company sample, which is in keeping with other studies of the UK company sector in similar periods (Dunne and Hughes 1992).

The final source of information concerning the scale of acquisition activity in the small company sector is drawn from a questionnaire completed by about 2,000 UK SMEs in 1991 (SBRC 1992). This questionnaire was designed as a follow-up and extension to the

Table 8.4 Death intensity amongst UK companies 1977–82 DTI Sample

| | | Total Assets 1976 | |
	Whole sample	≤ £1m	≥ £1m
Number of companies 1976	2,923	1,783	1,140
	(100%)	(100%)	(100%)
Acquisition deaths	200	63	137
1977–82	(6.8%)	(3.5%)	(12.0%)
Dissolved, receivership, etc.	604	496	108
1977–82	(20.7%)	(27.8%)	(8.5%)
Number of companies 1982	2,119	1,224	895
	(72.5%)	(68.7%)	(78.5%)

Note: 1 The sample is drawn from the DTI Company Accounts analysis and is based upon companies registered at Companies House in 1975. The sample is stratified with the following sampling proportions: 1 in 1 of the largest 500 companies; 1 in 2 of the next 500; 1 in 70 of medium sized companies; and 1 in 360 of the rest.

Table 8.5 Acquisition activity by firm size

Number of firms acquired in past 5 years	Whole sample
0	80.9%
1	12.5%
2	4.1%
3 or more	2.5%
Total	100%
Total responses (no.)	1,940
Total number acquired	600
Average number acquired all respondents	0.31
Acquirers only	1.62

Table 8.6 Acquisition intensity: comparison with Bolton manufacturing sample

	% of sample firms which have taken over or merged with other firms in the last five years	
	1990 SBRC sample	1969 Bolton sample
Manufacturing	20.3	8.8
Less than 100 employees	14.6	6.6
100–199 employees	27.9	18.4
Slow growth	14.6	5.4
Fast growth	21.3	16.0

Bolton enquiry of twenty years earlier (HMSO 1971). The behaviour of this sample as acquirers is summarized in Table 8.5 which shows that about 20 per cent of these SMEs had acquired another company within the previous five years. These 371 companies had between them acquired 600 companies in the previous five years and 6.6 per cent of the sample respondents had acquired more than one company. This scale of activity was more than twice as great as that observed by the Bolton enquiry (see Table 8.6). It must be remembered that 1986–9 was a period of high acquisition intensity, but so also was 1969 and so the comparison is roughly appropriate.

The questionnaire also enquired whether sample companies had been the subject of a takeover bid or merger proposal within the previous five years. These were necessarily bids which had not been completed since the sample contained only independent companies. The findings are shown in Table 8.7 and reveal that about a quarter of the sample had been the subject of a bid within the previous five years. The bids were received mainly from larger firms, with a surprisingly high proportion coming from foreign companies.

Table 8.7 Small firms as acquisition targets

	Whole sample
Sample size	2,023
% received bid in previous 5 years	23.0
Bid received from	
Larger firms, of which	20.7
UK company	14.1
Foreign company	3.6
Both	3.0
Firm of about same size, of which	4.3
UK company	3.6
Foreign company	0.5
Both	0.2

Therefore we find that acquisition activity is by no means the prerogative of large firms. Acquisitions within the small firm sector appear to follow a similar pattern over time to that observed for large firms, but this pattern is not even across industries. Small companies are less likely to be acquired, but more likely to fail than larger companies. Despite this there is a very active acquisition market within the small firm sector with a significant proportion of these companies being involved as acquirers, or targets, or both. Given the numerical dominance of the whole business sector by SMEs this means that the vast majority of acquirers and targets are smaller firms.

MOTIVATION AND ATTITUDES

Why do mergers occur? To answer this question requires a consideration of the motivation of both buyers and sellers. In the case of quoted companies whose shares are not closely held by their boards of directors the picture is complicated by the possibility of hostile bids in which one management team appeals directly to the shareholders of another company without the approval of, or in face of the opposition from, the target board. In the non-quoted sector with which we are primarily concerned sellers make their decision 'voluntarily'; control cannot be purchased against their will. Thus compared with quoted company takeovers we may ignore possible motives associated with the market for corporate control, in which inefficient non-owner managers are threatened with removal by a successful 'disciplinary' takeover bid. Why then might a buyer and seller in a world of closely held companies find acquisition a mutually attractive proposition?

The literature identifies several possible reasons. For the acquirer there are benefits from gains in market share and economies of scale, from diversification into new products or markets and from vertical integration. The acquired company may also be driven by these motives, but also by the prospect of capital gains from selling out or by the solution of management succession problems. These reasons were assessed by respondents to our questionnaire survey of 2,000 SMEs. Taking acquirers first, they were asked to assess the importance of a range of motives for acquisition and we were able to analyse the responses in terms of other characteristics of the acquiring company. The findings are summarized in Tables 8.8 and 8.9 which give the mean score given by our respondents to each motive on a scale of 0 (completed unimportant) to 9 (very important). Table 8.8 splits this sample into micro (< 10 employees), small (10–99 employees), medium (100–199 employees) and larger (200 ≤ 500) firms in order to assess whether any size effects exist. Not surprisingly, we find that capturing market share is identified as the most important motive. Diversification into new products and new areas are also important motives, along with the achievement of economies of scale. On the other hand, vertical integration is not generally important, perhaps because the suppliers and customers of our sample firms tend to be larger firms. It is difficult to identify a marked size effect on these motives. But we can suggest tentatively that vertical integration and gaining market share are more important amongst the micro firms and that economies-of-scale gains and product diversification are somewhat less important in this group.

Table 8.9 divides the sample first between manufacturing and

Table 8.8 Motives for acquisition by size groups (mean scores)

	Whole sample	Micro	Small	Medium	Larger
Horizontal					
Gain market share	5.8	6.4	5.6	5.2	6.1
Diversify					
New products	4.6	4.1	4.7	4.5	4.6
New geographical markets	3.7	4.0	3.5	3.5	4.0
Vertical					
Acquire supplier	1.3	1.8	1.3	1.2	1.2
Acquire customer	0.8	1.4	0.7	0.7	0.9
Efficiency					
Economies of scale	3.8	3.4	3.9	3.8	3.8
Number of acquirers	378	49	193	51	81

Table 8.9 Motives for acquisition by industry and growth rate (mean score)

	Manufacturing	Services	Stable/ Declining	Medium growth	Fast growth
Horizontal					
Gain market share	5.2	6.5	5.8	5.8	5.9
Diversify					
New products	4.6	4.5	4.0	4.8	4.5
New geographical markets	3.0	4.4	2.7	3.7	4.6
Vertical					
Acquire supplier	1.6	0.9	1.9	1.2	1.1
Acquire customer	1.0	0.7	0.8	0.8	0.7
Efficiency					
Economies of scale	3.6	4.1	3.2	4.0	4.0
Number of acquirers	210	168	77	158	87

business services and then by the growth experience of the respondent in the previous three years: stable/declining ($\leq$ 0 per cent), medium growth (0–75 per cent) and fast growth ($\geq$ 75 per cent), where the figures in parentheses represent employment growth 1987–90. It would appear that services acquirers are motivated more by market share, new geographical markets and, perhaps surprisingly, by economies of scale, and that manufacturing acquirers place somewhat greater emphasis on vertical integration. The fast growing acquirers appear to be driven more by new geographical and new product markets and by economies of scale, whereas the stable/declining group have been more likely to acquire suppliers. New and old firms were not distinguishable in terms of their motives for acquisition. This question allowed respondents to include other motives for acquisition and these were often revealing. The most frequently occurring reasons were: to acquire management skills or expertise; to reduce competition by acquiring a competitor; and to gain new premises or operating capacity. A handful of firms identified gaining access to new technology as an acquisition motive.

The questionnaire survey also asked firms about their attitude to being acquired. This question was asked in two parts – the first asked how firms felt about being taken over in the near future and the second asked for their longer-term attitude. In both cases more firms were opposed to the idea than were in favour of it, but a large proportion described themselves as open-minded on this issue. We found that the open-minded proportion rises from one-half in the near future to two-thirds in the longer term. The proportion opposed falls from 42 per cent to 20 per cent and the proportion in favour rises from 8 per cent to 14 per cent when considering the longer term in comparison with the near future. This supports those findings that suggest takeover to be an important and positive 'exit' route in small firm development for some owners, and is consistent with the small sample findings of Boswell (1972).

The sample findings show that a higher proportion are opposed and a lower proportion in favour of being acquired amongst medium-sized and larger companies than amongst small and micro companies, and that this is the case when considering either the near future or the longer term. Newer firms (and firms in the service sector) appear to be less opposed to being acquired than older firms, and this contrast is stronger in the longer term.

Whilst there is little to distinguish the findings in this area between stable/declining firms and medium growth firms, it is noticeable that fast growth firms have a more positive attitude to the question of being acquired in the longer term. This may be related to the fact that they are more likely to be micro or small companies.

The findings as a whole suggest that fast growing, small and medium-sized firms have a more open-minded or positive attitude towards being acquired. This may be related to its use as an exit route or means of realizing capital gains for founder managers (who are more frequently to be found amongst micro and small firms). On the other hand, it was also found in a separate analysis, not reported here, that companies with negative profits were less opposed to being acquired than others, and in this case we must suppose that survival is perhaps the driving force. It is worth noting here that results reported later in this chapter suggest that it is the less profitable acquired firms which gain most from takeover; and it is possible that their positive attitude may play some part in this success.

The sample firms were asked to assess a number of potential advantages of being acquired and scores have been calculated from their replies. The mean scores for the whole sample and the four size groups are shown in Table 8.10. For the whole sample the prospect of capital gains and increased market share are seen as the most important gains from being acquired. The next most important are the solution of management succession problems and the realization of economies of scale, followed by improvement in career prospects. The benefits of closer links to buyers or suppliers were not seen as important and this supports the findings of a low incidence of vertical integration acquisition behaviour in the small firm sector, which is in keeping with its insignificance amongst larger companies (Hughes 1993).

Table 8.10 Advantages of being acquired: by size group

| | Mean scores for potential advantages | | | | |
	Whole sample	Micro	Small	Medium	Larger
Prospect of capital gains	5.0	5.1	5.2	4.9	3.8
Solution to management succession	3.4	3.1	3.8	3.0	2.9
Improved career prospects	2.5	2.5	2.5	2.6	2.6
Increased market share	4.0	3.9	4.0	4.0	4.0
Gain economies of scale	3.2	2.8	3.2	3.7	3.7
Benefit from supplier links	1.7	1.9	1.6	1.3	1.4*
Benefit from buyer links	1.9	2.0	2.0	1.9	1.4
Total responses (no.)	1,858	496	1,015	169	161

The variation across size groups is not particularly marked, but the larger sample firms do not rate the prospect of capital gains as highly as other sample firms, nor do they rate highly the solution to the problem of management succession. We found that larger firms are predominantly companies and are also more likely to have relatively dispersed shareholdings and non-executive chairmen than smaller firms (SBRC 1992, Chapter 1). Such firms may already have made capital gains on their initial holdings and have made the transition of control from one generation to the next, or to professional managers. Takeovers may therefore hold less attraction for larger companies as a means of solving these problems. It is interesting to note that smaller firms see economies of scale as less important than larger firms, and this may reflect the fact that smaller firms are less likely to be competing on the basis of price (SBRC 1992). Whilst all groups rate the links with suppliers or buyers as not important, the smaller firms do score them somewhat higher.

Comparisons between manufacturing and service firms are shown in Table 8.11 and reveal that service firms view the advantages of capital gains and market share more highly than do manufacturing firms, which may be a reflection of their smaller size. This table also shows the findings for the sample split according to the firm's recent growth performance. It shows that the fast growth firms see the principal advantages of being taken over in terms of capital gains and market share – a similar finding to that for service firms. This suggests that in this area, at least, the service firms have a more dynamic attitude than do manufacturing firms, although the effect of size is also at work.

The sample companies were given the opportunity to suggest other advantages of being acquired and this identified three important aspects. The most frequently mentioned was the benefit of access to

Table 8.11 Advantages of being acquired: by industry and growth rate

	Mean scores for potential advantages				
	Manufacturing	*Services*	*Stable/ Declining*	*Medium growth*	*Fast growth*
Prospect of capital gains	4.9	5.2	4.5	4.9	5.8
Solution to management succession	3.6	3.3	3.5	3.8	3.1
Improved career prospects	2.3	2.6	2.3	2.6	2.6
Increased market share	3.7	4.3	3.7	4.0	4.1
Gain economies of scale	3.1	3.3	3.0	3.3	3.2
Benefit from supplier links	1.9	1.4	1.5	1.6	1.6
Benefit from buyer links	2.2	1.7	1.6	2.1	2.0
Total responses (no.)	977	877	476	684	392

financial resources and stability, a factor of great importance in the current economic climate and echoing the findings of recent case studies (ACOST 1990). The second was retirement of the founder, and can be taken as a particular form of the management succession problems. The third was the advantage of realizing owner investment, and can be associated with the answers relating to the prospect of capital gains. These results too mirror the findings of earlier UK studies and those of Ravenscraft and Scherer for the US. Another advantage suggested by several firms was the benefit of market diversification.

The sample firms were also asked to score how important to them would be the loss of independence and control if they were to be acquired. It is no surprise to find that firms generally scored this as a strong disadvantage. The mean score for all firms was 6.8 and there was little variation in this average across the different groups. As might be expected, the scores ranged from 7.9 for those who opposed being acquired in the near future, through 6.3 for the open-minded to 4.3 for those in favour, with similar, but lower, scores for the longer-term equivalents.

Firms were also asked to give other disadvantages of being acquired and many of these reinforced the point about loss of independence and control. Thus loss of motivation, dislike of the large company ethos, loss of flexibility, loss of focus, loss of identity and job prospects were all mentioned several times. On a more personal level, there were a number of mentions of the loss of employment opportunities for family members and the loss of face for the founders. The idea that 'running your own show' is an important aspect of small business motivation is strikingly confirmed by these responses.

SMALL FIRM ACQUIRERS AND TARGETS

The takeover wave of recent years has stimulated debate about the role played by acquisitions in creating efficiency at the level of the firm and for the economy as a whole. One aspect of this debate has involved an analysis of the characteristics of companies participating in acquisitions, and of the possible gains if acquired companies were notable under-performers whose link up with acquirers with superior performance led to improved performance. The attention of such research has focused almost exclusively on very large companies and this section makes a contribution towards redressing the balance. We shall assess whether the small-firm players in the acquisition process when compared with non-players show different relative charac-teristics from those found from equivalent studies of large-firm participants.

UK studies of large-firm acquirers have found them to be typically larger and faster-growing than their non-acquiring counterparts, but not consistently different in their other characteristics (Hughes 1993). We have analysed this issue by examining the characteristics of the acquirers in our questionnaire sample. The key findings are presented in Tables 8.12 and 8.13. We find a marked size effect which reinforces the findings of those studying the upper end of the size distribution of firms. Less than 10 per cent of the micro firms had acquired another company within the previous five years, but almost half of the larger sample respondents had done so. There is a clear pattern of increasing acquisition intensity with increasing size of firm. The

Table 8.12 Acquisition activity by firm size

	% of firms				
	Whole sample	Micro	Small	Medium	Larger
Numbers of firms acquired in past 5 years					
0	80.9	91.3	81.9	72.1	53.2
1	12.5	7.3	12.8	14.2	24.3
2	4.1	1.0	3.8	8.8	10.4
3 or more	2.5	0.4	1.5	4.9	4.6
Total	100%	100%	100%	100%	100%
Total responses (no.)	1,940	518	1,050	183	173
Total number acquired	600	67	274	87	166
Average number acquired					
all respondents	0.31	0.13	0.26	0.48	0.96
acquirers only	1.62	1.49	1.44	1.71	2.05

Table 8.13 Acquisition activity by firm growth

	% of firms		
	Stable/Declining	Medium growth	Fast growth
Number of firms acquired in past 5 years			
0	85.4	77.9	78.7
1	9.5	15.0	12.9
2	3.3	4.7	4.5
3 or more	1.8	2.4	3.9
Total	100%	100%	100%
Total responses (no.)	506	707	403
Total number acquired	112	235	157
Average number acquired			
all respondents	0.22	0.33	0.39
acquirers only	1.51	1.51	1.83

table shows that larger firms are also more likely to have acquired more than once.

An interesting question is whether acquisition is used as an important source of growth for small firms. We can ask whether fast growth firms use acquisitions more intensively. This issue is addressed in Table 8.13 which reveals a number of points. Small firms which are acquirers are found to be more dynamic than their non-acquiring counterparts and this again supports the findings of large-firm studies. Further analysis reveals that whilst the acquisition propensity of fast growers is greater than those which are stable or declining in size, it is not higher than medium-growth firms. But the bottom row of the table shows that fast growth acquirers have acquired more firms on average than the medium-growth acquirers, that is to say that they are more intensive acquirers. This supports the findings of a longitudinal study of SMEs in the London area in the 1980s (Smallbone, North and Leigh 1992). However, it is worth noting that about 80 per cent of the fast-growth sample firms have not used acquisition, but instead have achieved these high growth rates through organic growth. This is probably a reflection of the fact that fast-growth firms are much more likely to be micro and small firms. Finally, it is worth noting that a separate analysis not reported here suggested there was no relation between acquisition intensity and the profitability of the firm – a result which is to be found in most studies of large companies.

Before turning to examine the characteristics of a sample of acquired firms we can draw upon the questionnaire sample of acquisition targets. The characteristics of these targets are shown in

Table 8.14 Small firms as acquisition targets

	Whole sample	Size effect			Age effect			Growth effect		
		Micro	Small	Medium	Larger	Older	Newer	Stable/Declining	Medium growth	Fast growth
% received bid in previous 5 years	23.0	14.8	24.7	32.2	29.0	24.9	21.0	21.8	25.9	26.5
Bid received from Larger firm	20.7	12.8	22.0	30.6	27.3	23.1	18.0	19.4	24.1	23.9
Firm of about same size	4.3	2.2	4.7	7.1	5.1	4.4	4.1	3.9	5.4	4.6

Source: SBRC Report Table 4.9

Table 8.14 analysed by size, age and growth. The effect of firm size is again quite marked, with the proportion of firms who had received an approach rising from the smallest to largest size groups. Whilst 12.8 per cent of micro firms had received a bid from a larger company in the previous five years, the proportion rose to 30.6 per cent for medium-sized firms. In a similar fashion, bids from firms of about the same size were received by 2.2 per cent of micro firms and by 7.1 per cent of medium-sized firms. It is worth noting that there is some evidence of a reduction in the proportion of larger firms who had received a bid in comparison with medium-sized firms. Whilst one could speculate whether this suggests that there is an optimum target size amongst the small firm sector (perhaps where we might expect financial and managerial constraints to be greatest), the evidence here cannot be regarded as conclusive. We do find, however, that foreign firms are more likely to be interested in the larger members of our sample as acquisition targets.

Old firms were more likely (23 per cent) to have received an approach from larger companies than were newer firms (18 per cent), and in the case of older firms a higher proportion of such approaches came from foreign firms. Turning to the question of whether the growth performance of the sample firms had any effect on their probability of receiving a bid, the only difference worth noting is that the stable/declining sample were somewhat less likely (19 per cent) to have received bid from a larger firm than the other two growth groups (24 per cent).

Previous studies of acquired firms in the large-firm sector have found them to be smaller and slower growing than non-acquired firms. The evidence on profitability is mixed but there is some support for the view that acquired companies have a worse profitability record just prior to acquisition. These findings based on large-firm samples are not inconsistent with the above analysis of the questionnaire sample targets. However these are *target* companies where the bid was not effective and so cannot take them to be characteristic of acquired companies.

To carry the analysis further, a sample of acquired companies was drawn from the DTI company sample which was described earlier and which covers the whole corporate sector. These sample companies were matched by size and industry with other companies which had not been acquired. In this way we can distinguish the financial characteristics of acquired companies from other companies after correcting for size and industry effects. The financial characteristics shown in Table 8.15 measure liquidity and financial structure, borrowing, growth and profitability. The findings are also presented separately for small and large acquired companies, where total assets of £1 million is used as the dividing line.

Looking first at the liquidity measures we find somewhat lower ratios of net current assets to total assets for acquired companies and that this is more marked when considering the median. We also find that this ratio is smaller for small companies, but none of these differences is statistically significant. The acid test ratio which measures the ratio of current assets less stocks to current liabilities introduces two changes to this pattern. First, the difference between large and small firms is less. Second, the median for the acquired sample is significantly less than for their matches and this finding can be seen to be due to the lower median value of this liquidity measure for large acquired firms than for the large matched. This pattern is repeated for the net liquid asset ratio, but none of the differences are significant. The final measure of liquidity is the ratio of net current assets to net assets which again shows no significant differences either between large and small acquired companies or between these groups and their matched samples. The ratio of fixed assets to total assets is significantly less for small acquired than large acquired companies, but this is clearly due to the size effect since neither is significantly different from their matched samples.

The three measures of gearing provided are short-term loans, long-term loans, and all loans as a proportion of total assets. The short-term measure shows a tendency for acquired companies to have a higher gearing. For this sample this is more pronounced amongst the large acquired sample where the differences from its matched sample are statistically significant. This picture is reversed in terms of long-term borrowing since it is the small, acquired firms which have a much lower gearing than the other groups. Taking short-term and long-term gearing together we find only small differences between acquired companies and their matches, but do find higher overall gearing for larger companies in comparison with the small companies.

The findings for profitability and growth show more significant differences. The ratios of net profits to net assets and to total assets show the acquired company sample to be less profitable than the matched sample in the year prior to acquisition. This difference is statistically significant for the difference in means for all acquired and for large acquired. The third measure adds back total directors emoluments to profits before forming the ratio with total assets. This is particularly relevant for closely held companies where the emoluments of directors is an alternative means of distributing profits to

Table 8.15 Pre-acquisition characteristics of acquired companies (DTI sample, 1977–82, average values in year prior to acquisition)

		Whole sample (142 companies)		Small (35 companies)		Large (107 companies)	
		Acquired Sample	Matched Sample	Acquired Sample	Matched Sample	Acquired Sample	Matched Sample
Liquidity and Structure							
Current assets	Mean	1.56	1.66	1.35	1.59	1.63	1.69
/current liabilities	Median	1.32	1.50	1.11	1.17	1.43	1.60
Current assets – stocks	Mean	0.93	1.01	0.90	1.02	0.94	1.01
/current liabilities	Median	0.68[1]	0.83	0.69	0.70	0.68[2]	0.87
Net liquid assets	Mean	−0.08	−0.06	−0.07	−0.07	−0.08	−0.05
/total assets	Median	−0.10	−0.06	−0.11	−0.09	−0.09	−0.04
(Current assets–current liabilities)	Mean	0.35	0.32	0.40	0.21	0.34	0.35
/Net assets	Median	0.38	0.39	0.49	0.41	0.37	0.39
Fixed assets	Mean	0.36	0.37	0.28	0.32	0.38	0.38
/total assets	Median	0.31	0.31	0.21	0.20	0.34	0.32
Borrowing							
Short-term loans	Mean	0.12	0.11	0.10	0.12	0.13[1]	0.10
/total assets	Median	0.11[2]	0.07	0.09	0.08	0.12[1]	0.07
Long-term loans	Mean	0.05[2]	0.07	0.01	0.05	0.06	0.08
/total assets	Median	0.02	0.03	0.00	0.00	0.03	0.04
All loans	Mean	0.18	0.18	0.11	0.17	0.20	0.18
/total assets	Median	0.17	0.15	0.09	0.12	0.20	0.16

Table 8.15 continued

		Whole sample (142 companies)		Small (35 companies)		Large (107 companies)	
		Acquired Sample	Matched Sample	Acquired Sample	Matched Sample	Acquired Sample	Matched Sample
Profitability and Growth							
Net profit	Mean	0.12[1]	0.18	0.17	0.21	0.11[1]	0.17
/ave net assets	Median	0.14	0.17	0.15	0.20	0.14	0.17
Net profit	Mean	0.06[2]	0.09	0.06	0.05	0.06[1]	0.10
/ave total assets	Median	0.07	0.09	0.05	0.08	0.07	0.10
Net profit + dir emols	Mean	0.08[1]	0.11	0.15	0.17	0.07[1]	0.10
/ave total assets	Median	0.09	0.11	0.18	0.14	0.08	0.10
Growth of	Mean	0.08	0.13	0.12	0.16	0.07[1]	0.13
total assets	Median	0.10	0.12	0.10	0.11	0.10	0.12

Notes: The sample of 142 acquired companies was taken from the full sample of acquisition deaths shown in Table 8.4 by excluding those: with total assets less than £50,000 in 1976, or with insufficient data, or those which were 'born' and acquired within the same period. Small companies were defined as those with total assets of less than £2.5m in 1976.

1 indicates that the average for the acquired company sample is significantly different from the average for the matched sample at the 5 % level.

2 indicates that the average for the acquired company ample is significantly different from the average for the matched sample at the 10 % level.

the owners. In terms of this measure the small acquired sample is significantly more profitable than the large acquired sample. Furthermore the large acquired are significantly less profitable on average than their matched companies. Finally the growth measure shows that the acquired companies were also slower growing on average than their matched counterparts immediately prior to acquisition and the differences in means are significant for the large acquired group.

For small acquired companies as a group we may summarize our results as showing them to be somewhat less profitable, slower growing, less liquid and with higher short-term borrowing than those not acquired. They are therefore underperformed with signs of financial distress.

ACQUISITION EFFECTS

Traditional approaches to the assessment of acquisition effects have involved share price event studies and accounting-based research. The share price approach is not available for studies of the effects of acquisition on small unquoted companies and there are significant problems with the accounting-based methodology. The accounting-based studies of acquisition effects using samples of larger companies compare the actual post-acquisition performance of the merged companies with their asset weighted combined pre-merger performance. This works well when the acquired company is reasonably large in relation to the acquirer. In such cases one can expect that the impact of acquisition on the performance of the acquirer will be sufficiently significant not to be lost in the noise of other events. But when we are studying the acquisition of small firms this is not a reasonable expectation and an alternative approach is needed.

Fortunately an alternative method is available in some cases since acquisition does not always result in a cessation of financial reporting by the acquired company. From our DTI sample containing 200 acquisition deaths in the period 1977–82 it was possible to select a group of such companies. Our requirements were that we had accounting information for the acquired company in the two years prior to acquisition and for at least three years after acquisition (data for the year of acquisition was excluded). The accounts of such companies were carefully checked to see whether they appeared to exhibit any significant changes in accounting treatment. This approach yielded a sample of forty companies which were then analysed.

This sample was first checked against the whole sample of acquisition deaths to check for selection bias. We found that the reduced sample was representative of the full sample in terms of both size and industry. The characteristics of the acquired firms are reported for the two years prior to acquisition in Table 8.16. The table shows the mean and median values for each variable adjusted relative to the industry (calculated by subtracting from the value of that variable for each acquired company the value of the industry median and then calculating the mean or median of these adjusted variables). In this way each characteristic of the acquired companies is measured relative to industry average performance. If the mean of these observations were zero it would suggest that acquired firms were on average no different from other firms in their industries in terms of that variable.

If we look at the first column of results shown in Table 8.16 we can see the findings for the two years prior to acquisition for the sample of forty companies as a whole. This shows that the acquired firms are somewhat smaller on average than the average for their industries but the median difference is quite small. Most of these comparisons reveal insignificant differences between the acquired sample and their industry means and are broadly in line with the findings for the full sample presented in Table 8.15. However whilst the profitability of the full DTI sample was found to be less than their matched counterparts, the difference is smaller and not statistically significant here. This may be partly due to the smaller sample size and to using the industry median rather than a matched sample for our comparison. The most significant difference is the finding for this group of acquired firms that they had higher ratios of long-term loans to total assets in the two years prior to acquisition than did the average firm in their industry.

This sample is further partitioned in two ways and the next two columns show it divided by the size of the acquired company. This shows some marked differences between small and large acquired companies (divided by total assets greater or less than £1m). The small exhibit significantly lower liquidity, assets to sales ratios and profit margins. On the other hand when profitability is measured by the return on net assets it is the large acquired sample which is significantly less profitable on average. In fact the small acquired firms in this sample are on average more profitable than their industries. This was not found for the larger sample reported in Table 8.11. The Table 8.11 sample did however show the large acquired to be relatively less profitable than the small acquired, which is also true in the sample under analysis here.

This sample was also partitioned by the success of the acquisition.

Table 8.16 Pre-acquisition characteristics of acquired companies

Variable	Average Measure	Average values in two years prior to acquisition (Relative to their industry medians)				
		All	Large	Small	Merger Success	Merger Failure
Total assets	Mean	26,561[1]	47,588[1]	-8,485[1]	44,778[1]	11,656
	Median	481	24,414[1]	-7,177[1]	19,585	-420
Sales	Mean	34,944[1]	57,916[1]	-12,915[1]	52,946[1]	21,229
	Median	-2,048	22,252	-13,258[1]	27,486	-4,818
Current assets /current liabilities	Mean	0.03	0.14	-0.16	0.01	0.05
	Median	-0.13	0.10	-0.19[2]	-0.14	-0.05
(Curr assets-curr liabilities) /net assets	Mean	-0.08	-0.02	-0.17	-0.16[2]	-0.01
	Median	-0.03	0.05	-0.11	-0.14	0.05
Fixed assets /sales	Mean	0.04	0.09[1]	-0.05	0.06[2]	0.03
	Median	-0.01	0.04	-0.06[1]	0.03	-0.03
Net assets /sales	Mean	0.02	0.10[1]	-0.14[1]	0.04	0.01
	Median	0.01	0.06[1]	-0.15[2]	0.06	0.01
Long-term loans /total assets	Mean	0.046[1]	0.048[1]	0.042[1]	0.044[1]	0.047
	Median	0.033[1]	0.034[1]	0.005	0.025	0.034
Net profit /ave total assets	Mean	-0.012	-0.015	-0.007	-0.071[1]	0.036[1]
	Median	0.004	0.005	0.003	-0.072[2]	0.025[2]
Trading profit /sales	Mean	-0.007	0.002	-0.026[1]	-0.037[1]	0.016
	Median	0.004	0.007	-0.020	-0.025	0.011
Equity return	Mean	0.018	-0.038	0.114	-0.113[1]	0.124[1]
	Median	0.016	0.016	0.023	-0.030	0.065[1]
Growth of total assets	Mean	0.01	-0.02	0.06	-0.09[1]	0.09[1]
	Median	-0.02	-0.03	0.09	-0.08[1]	0.06[2]

Table 8.16 continued

| Average Variable | Measure | Average values in two years prior to acquisition (Relative to their industry medians) | | | |
		All	Large	Small	Merger Success	Merger Failure
Net income	Mean	0.00	−0.06	0.10	−0.16[1]	0.13[2]
/ave net assets	Median	−0.01	−0.01	0.02	−0.12[1]	0.04
Number of companies		40	25	15	18	22

Notes: For definitions of variables and meaning of Large, small, merger success and merger failure see text.
1 indicates significantly different from industry average at the 5% level
2 indicates significantly different from industry average at the 10% level.

Success was measured by the change in industry normalized profit-ability from two years before to three years after the acquisition. If this change was positive the acquired firm was put in the 'merger success' sample. What emerges most clearly is that the successful acquisitions involve turnround of performance from a significant underperformance in terms of profitability. The successful acquisitions were associated with acquired companies with lower profit-ability than their industries. On the other hand, acquired firms with above average profitability tended to show declines in profitability as a consequence of the acquisition. The 'merger failure' sample were also superior performers in terms of pre-acquisition growth.

The impact of acquisition on relative profitability is examined further in Table 8.17. The first row concerns the whole sample and reveals that whilst relative profitability was above industry average prior to acquisition, it is slightly below average post-acquisition. This finding of an insignificant deterioration of profitability performance has been found in most large company merger studies. However, a more encouraging picture emerges when we look five years beyond the acquisition event. Lack of data restricted this sample to twenty-three companies and the results for these are shown in the second row of Table 8.17. They show a similar profitability performance in the first three years post-acquisition relative to their industries. However since this group had a better pre-acquisition profitability than the whole sample, the decline in profitability over this period is more marked for the 5-year sample. Most interestingly there is a sub-stantial improvement in relative profitability in the fourth and fifth years post-acquisition. This finding is supported when we partition the sample. Small acquired companies show a much greater deteriora-tion in profitability in the three and five years post-acquisition than do larger acquired companies. But in both cases the relative profitability performance improves towards the end of the five year period.

Finally the bottom rows of Table 8.17 show a comparison of the successful and unsuccessful groups. The most striking finding is that the successful group were those companies which brought their profitability from below industry average in the two years prior to acquisition up to industry average in the three years after. Therefore acquisition appears to have the best chance of improving relative profitability when the acquired company was performing poorly relative to its industry prior to acquisition. Interestingly, this corre-sponds with our survey finding that the least profitable firms were most likely to view the prospect of being acquired with an open or positive mind. The unsuccessful group is dominated by companies

Table 8.17 Profitability performance of acquired companies – before and after takeover

Group	Coverage	Sample size	Two years pre-merger	Three years post-merger	Five years post-merger	Three years post -two years pre-merger	Five years post -two years pre-merger
Whole sample	All	40	0.00	-0.05	–	-0.05	–
	5 year	23	0.09	-0.06	-0.01	-0.14[1]	-0.10[2]
Small sample	All	15	0.10	-0.05	–	-0.15[2]	–
	5 year	9	0.22	-0.06	0.01	-0.28[2]	-0.21
Large sample	All	25	-0.06	-0.05	–	0.01	–
	5 year	14	0.00	-0.06	-0.03	-0.05	-0.03
Merger success	All	18	-0.16[1]	0.00	–	0.16[1]	–
	5 year	8	-0.11[1]	-0.01	-0.04	0.09	0.06
Merger failure	All	22	0.13[2]	-0.09	–	-0.22[1]	–
	5 year	15	0.19[2]	-0.08	0.00	-0.27[1]	-0.18[1]

Notes: Mean values for profitability measured relative to the industry median. Profitability is measured by the ratio of net income to net assets.
1 indicates significance at the 5% level.
2 indicates significance at the 10% level.

who were doing well relative to their industries before being acquired and whose performance post-acquisition is dismal. In this case we again observe an improvement in relative profitability five years after acquisition in comparison with the first three years post-acquisition.

The detailed information for this sample was drawn from microfiches of the companies' filings at Companies House. This enabled us to also identify the consequences of acquisition for the Board of Directors. The pre-acquisition characteristics of and the takeover effects on the Board of Directors are shown in Table 8.18. This information is presented both for the whole sample and for small companies within the sample. The upper half of this table reveals as we might expect that the board size was smaller, board shareholdings were greater and the average age of directors was somewhat less for the small companies. The acquisition event appears to have had little impact on the resulting board size, but does show that on average there is an extended period over which pre-acquisition board members are replaced. Continuity is achieved by retaining at least some of the original directors even five years after the takeover. It is quite possible that this finding is not characteristic of the typical acquisition. This sample consists of those acquired companies which

Table 8.18 Board of Directors – analysis of 40 acquired companies (acquired between 1977 and 1982)

	Whole sample mean values	*Small companies mean values*
Pre-takeover characteristics		
Number of directors	6.3	4.0
Largest % of shareholding	16.5%	31.2%
Board % shareholding	27.5%	49.1%
Average age of directors	56 years	51 years
Youngest director	44 years	34 years
Oldest director	68 years	65 years
Takeover effects		
Number of new directors appointed initially	2.1	1.7
Number of pre-takeover directors continuing after:		
1 years	3.2	2.0
3 years	2.2	1.4
5 years	1.5	0.6
Final number of directors	6.4	3.9
Number of companies	40	15

Note: For the purpose of this analysis small companies are taken to be those with net assets ⩽ £1m.

continued to report their results as an independent unit for at least three years after the acquisition. This might reflect a greater autonomy from the parent company than usual and therefore make it more likely that pre-takeover directors are retained.

In summary of this section we can compare our findings with those of Ravenscraft and Scherer. Their sample of acquired firms differed from ours in that their average profitability was above industry norms. On the other hand, they too found that smaller targets were more profitable than larger targets. In terms of the effects of acquisition on profitability they separated their sample into those which were subsequently divested and those which survived. The former group were abject failures. The targets were often acquired at, or near, their profitability peak and declined substantially thereafter. In this way they may be compared with our merger failure group which showed a similar pattern. However it is the survivors group with which our findings should be compared. For this group Ravenscraft and Scherer found a profitability performance post-acquisition in line with industry norms. This represented a significant decline from their pre-acquisition levels; and, as with our study, they found the decline in profitability to be greater for smaller firms. They attribute part of the decline to new and complex organizational structures being imposed on the targets and to loss of motivation. 'Although some of the decline is attributable to the unsustainably high level of pre-merger profits, an appreciable fraction appears to be a scaled-down manifestation of the control loss problems that led to sell-off in more extreme cases'. (Ravenscraft and Scherer 1987: 192–3). Our findings are in accordance with this conclusion, but we keep an open mind at this stage about the relative importance of the acquisition event itself or, alternatively, some form of Galtonian regression effect in which high profits converge to the mean (Cosh, Hughes, Lee and Singh 1989) in bringing them about.

SUMMARY

This chapter has presented the first systematic attempt to document and assess the scale, determinants and effects of acquisition activity involving small firms. It should provide a starting point for research on takeover activity in the SME sector. Until further work is carried out the findings presented here must be regarded as tentative. The chapter however has demonstrated that the scale of this activity makes it an important subject for further research. The principal findings may be summarized as follows:

1 In terms of the scale of activity we find that acquisitions amongst the SME sector numerically dominate overall acquisition activity and follow a similar time series pattern to that observed amongst larger companies.

2 The death rate amongst SMEs is larger than that for larger companies. For SMEs acquisition deaths represent about one-quarter of total deaths, but this proportion is much greater for larger companies. Overall SMEs have a lower probability of being acquired than larger companies.

3 About one-fifth of SMEs had acquired at least one other company and about one-quarter had been in receipt of an acquisition proposal in the period 1986–91.

4 SME acquirers scored capturing market shares, diversification into new products and new markets and gaining economies of scale as their most important motives for acquisition.

5 SMEs value their independence greatly and are concerned that acquisition would bring with it loss of motivation and flexibility. For these reasons more small firms opposed the ideas of being acquired than favoured it, but the majority were open-minded about the proposal in the longer term.

6 SME acquirers tend to be larger and faster growing, but not more profitable than non-acquiring SMEs.

7 SMEs are more likely to be acquisition targets if they are older and are amongst the larger SME size classes.

8 Acquired SMEs have on average lower profitability, somewhat lower growth, lower liquidity and higher short-term borrowing than non-acquired SMEs.

9 Acquisition of SMEs in our sample appears to have had a negative impact on average in terms of profitability performance, but there is some evidence that the deterioration is substantially reduced by five years after the acquisition. We also find that those companies which were relatively profitable prior to acquisition exhibited the greatest decline in relative profitability following the acquisition.

These findings show many more similarities than differences in the acquisition process amongst SMEs compared with large companies. This is something of a surprise since the nature of the process differs substantially between the groups. One might have expected that the lack of stock market quotations, the tightness of shareholdings, the involvement of owner-managers and differences in the way in which seller and buyer are brought together would result in significant differences in the relative characteristics of acquired and acquiring

314   *Andy Cosh and Alan Hughes*

companies and in post-merger effects. This, it appears on the evidence so far, is not in general the case, although further work is needed before we can be fully confident about this discovery and its causes.

NOTES

This paper arises from the research programme into the Determinants of the Birth, Growth and Survival of Small Businesses at the Small Business Research Centre (SBRC), Cambridge University. The research was supported under the ESRC Small Business Programme by contributions from the ESRC, Barclays Bank, Commission of the European Communities (DG XXIII), Department of Employment and the Rural Development Commission. This support is gratefully acknowledged. The authors are grateful to Simon James, Una Kambhampati, Marc Taylor and Mark Wilson for valuable research experience.

1 Ashcroft, Love and Scouller (1987) provide an interesting analysis of takeovers in Scotland with a sample of smaller firms, but these are mainly all in excess of the SME employment cut offs of 200 or 500 employees.
2 Birley and Westhead (1988) provide a useful analysis of numbers of businesses listed for sale, including those having ceased to trade, in the national and provincial press in the period 1983–7 and show an increasing trend in numbers.

REFERENCES

Advisory Council on Science and Technology (ACOST) (1990) *The Enterprise Challenge: Overcoming Barriers to Growth in Small Firms*, HMSO, London.
Ashcroft, B.K., Love, J.H. and Scouller, J. (1987) 'The economic effects of the inward acquisition of Scottish manufacturing companies 1965–80', *ESG Research Paper no. 11*, Industry Department for Scotland.
Auerbach, A.J. (ed.) (1988) *Corporate Takeovers: Causes and Consequences*, NBER, University of Chicago Press, Chicago.
Barber, J., Metcalfe, S.J. and Porteous, M. (eds) (1989) *Barriers to Growth in Small Firms*, Routledge, London.
Birley, S. and Westhead, P. (1988) 'Exit Routes' Mimeo, Cranfield School of Management, Cranfield.
Boswell, J. (1972) *The Rise and Decline of Small Firms*, Allen & Unwin, London.
Cosh, A.D., Hughes, A., Lee, K.C. and Singh, A. (1989) 'Predicting Success: pre-merger characteristics and post-merger performance', Small Business Research Centre Working Paper, no. 6, Department of Applied Economics, University of Cambridge.
Dunne, P. and Hughes, A. (1992) 'Age, Growth and Survival Revisited, Small Business Research Centre Working Paper, no. 23, Department of Applied Economics, University of Cambridge and (1994) *Journal of Industrial Economics,* June.

Garnsey, E. and Roberts, J. (1990) 'Growth through acquisition for small high technology firms. A case comparison' in S. Birley (ed.) *Building European Ventures*, Elsevier, Amsterdam.

Garnsey, E., Alford, H. and Roberts, J. (1990) 'Acquisition as long term venture: cases from high technology industry' *Small Business Research Centre Working Paper*, no. 9, Department of Applied Economics, University of Cambridge.

Hughes, A. (1989) 'Small firms' merger activity and competition policy' in Barber, J., Metcalf, S.J. and Porteous, M. (eds) *Barriers to Growth in Small Firms*, Routledge, London.

Hughes, A. (1993) 'Mergers and economic performance in the UK. A survey of the empirical evidence 1950–1990', in Bishop, M. and Kay, J.A. (eds) *European Mergers and Merger Policy*, Oxford University Press, Oxford.

Merret Cyriax Associates (1971) *Dynamics of Small Firms*, Research Report no. 12, Committee of Inquiry on Small Firms, London, HMSO.

Ravenscraft, D.J. and Scherer, F.M. (1987) *Mergers, Sell-offs, and Economic Efficiency*, Washington Brookings Institution.

SBRC (1992) *The State of British Enterprise: Growth, Innovation and Competitive Advantage in Small and Medium-sized Firms*, Small Business Research Centre, Cambridge.

Smallbone, D., North, D. and Leigh, R. (1992) 'Managing change for growth and survival: a study of mature manufacturing firms in London during the 1980s'. *Working Paper no. 3*, Planning Research Centre, Middlesex University.

Index

accounting standards *see*
management
accounting
acquisitions and mergers: effects 12,
287, 305–12; motivation and
attitudes 11–12, 56, 285–7, 292–7;
scale of activity 287–92; small firm
acquirers and targets 13, 298–305;
US study 284–5
advertising 209, 228
Advisory Council on Science and
Technology (ACOST) 2, 6, 28,
55–7, 68–70, 112, 121, 155, 284,
297
Afro-Caribbean community: access
to capital 145; business formation
difficulties 8, 13, 147–8;
comparison with Asian
community 147–50; funding 151,
153, 173–4, 176–7; informal
sources of finance 153; personal
savings 156, 158–9; post-start up
funding 174–6; relationship with
banks 8, 150–1, 153, 164–9; role of
family and friends 159–64; self-
employment rate 147
Aghion, P. 33
Albach, H. 113, 115–16
Aldrich, H. 154, 163
Alford, H. 284
Allen, K.R. 66–7
Altman, E. 29
Ang, J. 20
angels, business: characteristics 4–5,
70, 76–80; disadvantages 101;
'finance gap' debate 12, 68–9; high
technology sector 7, 86–7;

involvement 70; motivations 100;
referral networks 84–6, 101–2,
103–4; risk perceptions 95–7; role
100; satisfaction 98–9; survey
methodology 71–6; taxation issues
104; US 2, 4–5, 68–9, 72, 76,
92–3, 97, 99; 'virgin' 102–3, 104;
see also investment, informal
Aram, J. 72, 76, 105 n. 2
Aram Associates 72, 105 n. 2
Ashcroft, B.K. 323 n. 1
Asian community: access to capital
145; business funding 150–3,
169–73, 176–7; businesses 2, 8, 13,
146–50; comparison with Afro-
Caribbean community 147–50;
informal sources of finance 153–4;
personal savings 156–8; post-start
up funding 174–6; relationships
with banks 150–3, 164–9; role of
family and friends 159–64; self-
employment rate 146–7
Asian Business 147, 164, 166
asset structure 34–6
Aston, University of 55
Auerbach, A.J. 283
Aunt Agatha 54

Bains, H. 149
Ballard, C. 149
Ballard, R. 149
Bank of Credit and Commerce
International (BCCI) 151–2
Bank of England (1993) 2
Bank of International Settlements 64
banks: attitude to incorporation
266–7; credit rationing 9, 53–4, 66;

Gaston, R.J.: (1989a) 68, 72; (1989b) 76–7, 79, 88, 90, 93, 96, 98–9; and Bell (1986) 72, 105, n.2; and Bell (1988) 72
Godwin, M. 10–11, 14, 233, 243, 250, 279 n. 6
Goffee, R. 27, 155
Goodman, J.P. 66–7
Government of Ontario 69
government role 14, 119–20; DTI 6, 104
Gower, L.C.B. 237, 240–1, 278 nn. 5, 13
Grant Thornton 67
Gray, C. 27, 182, 245
Greenfield, S. 9, 13, 14
Gretton, E. 160
Groves, R. 37
growth rates 27

Haar, N.E. 71–2, 76, 105 n. 2
Hakim, C. 27, 234, 256, 267, 279 n. 14
Hall, G. 120
Halpern, P. 242, 262, 266
Hankinson, A. 184
Hansmann, H. 279 n. 7
Harman, J. 239
Harrison, R.T.: and Mason (1991) 105 n. 3; and Mason (1992a) 88, 104; and Mason (1992b) 5, 89, 91, 100; Groves and (1974) 37; Mason and (1991) 100; Mason and (1992) 68, 101; Mason and (1993a) 105 n. 2, 106 n. 10; Mason and (1993b) 103; Mason and (1993c) 104; Mason and (1993d) 104; Mason and (this volume) 4–6, 12, 15
Hart, O. 33
Harvard Securities 75
Hauptman, O. 114
Helweg, A. 149, 156
high technology sector: external sources of finance for expansion 7, 136–7; finance constraints and age of company 13, 124–5; finance constraints and growth 13, 125–7; finance constraints by sector 13, 121–4; importance of finance constraints 6–7, 132–4; start up finance 7, 113–21

hire purchase 32, 33, 51
Hirsch, R.S. 27
Hiro, D. 149
hi-tech *see* high technology
HMSO (1971) (Bolton Report) 1, 2, 3, 18, 21, 26, 34, 37, 42, 50–1, 112, 136, 182, 279 nn. 14, 24, 281 n. 26, 292
HMSO (1979) (Wilson Report) 1, 2, 3, 18, 34, 37, 42, 53–5, 112
HMSO (1991) 2, 32, 55
Holmes, S. 37, 184
Holström, B. 30
Home Office 155
Hornaday, R.W. 27
Hudson, J. 240, 269
Hughes, A.: (1991) 284, 285, 295, 298; (1992) 56–7; Cosh and (1989) 21, 24; Cosh and (1994) 49; Cosh and (this volume) 3–4, 11–12, 13, 15, 27, 37, 41, 45; Cosh, Hughes and Singh (1990) 25; (1992) 1, 11, 14; Cosh, Hughes, Lee and Singh (1989) 312; Dunne and (1994) 290
Hunsdiek, D. 113, 115–16
Hussain, G. 64
Hutchinson, O. 37
Hutchinson, R.W. 112
'hybrid workers' 255–8

ICAEW (1992) 250, 274
ICAEW (1993) 252
Income and Corporation Taxes Act (1988) 247, 256, 279 n. 15
incorporation: advantages 249–50; attitudes to 232–3, 265–7; boom 10–11; capital factor 258–62; disadvantages 250–5; disincorporation 252–4; economic literature perspectives 234–6; costs and benefits 242–3; 'hybrid workers' 255–8; legal perspective 236–42; minimum capital requirement 11, 14, 268–9; reasons for 245–7; share capital 262–5; small business literature perspective 234; survey 10–11, 243–65; taxation 247–9
informal finance *see* investment (informal), investor (informal)